Lecture Notes in Physics

Springer-Verlag Berlin Heidelberg GmbH

The Editorial Policy for Proceedings

The series Lecture Notes in Physics reports new developments in physical research and teaching – quickly, informally, and at a high level. The proceedings to be considered for publication in this series should be limited to only a few areas of research, and these should be closely related to each other. The contributions should be of a high standard and should avoid lengthy redraftings of papers already published or about to be published elsewhere. As a whole, the proceedings should aim for a balanced presentation of the theme of the conference including a description of the techniques used and enough motivation for a broad readership. It should not be assumed that the published proceedings must reflect the conference in its entirety. (A listing or abstracts of papers presented at the meeting but not included in the proceedings could be added as an appendix.)

When applying for publication in the series Lecture Notes in Physics the volume's editor(s) should submit sufficient material to enable the series editors and their referees to make a fairly accurate evaluation (e.g. a complete list of speakers and titles of papers to be presented and abstracts). If, based on this information, the proceedings are (tentatively) accepted, the volume's editor(s), whose name(s) will appear on the title pages, should select the papers suitable for publication and have them refereed (as for a journal) when appropriate. As a rule discussions will not be accepted. The series editors and Springer-Verlag will normally not interfere with the detailed editing except in fairly obvious cases or on technical matters.

Final acceptance is expressed by the series editor in charge, in consultation with Springer-Verlag only after receiving the complete manuscript. It might help to send a copy of the authors' manuscripts in advance to the editor in charge to discuss possible revisions with him. As a general rule, the series editor will confirm his tentative acceptance if the final manuscript corresponds to the original concept discussed, if the quality of the contribution meets the requirements of the series, and if the final size of the manuscript does not greatly exceed the number of pages originally agreed upon. The manuscript should be forwarded to Springer-Verlag shortly after the meeting. In cases of extreme delay (more than six months after the conference) the series editors will check once more the timeliness of the papers. Therefore, the volume's editor(s) should establish strict deadlines, or collect the articles during the conference and have them revised on the spot. If a delay is unavoidable, one should encourage the authors to update their contributions if appropriate. The editors of proceedings are strongly advised to inform contributors about these points at an early stage.

The final manuscript should contain a table of contents and an informative introduction accessible also to readers not particularly familiar with the topic of the conference. The contributions should be in English. The volume's editor(s) should check the contributions for the correct use of language. At Springer-Verlag only the prefaces will be checked by a copy-editor for language and style. Grave linguistic or technical shortcomings may lead to the rejection of contributions by the series editors. A conference report should not exceed a total of 500 pages. Keeping the size within this bound should be achieved by a stricter selection of articles and not by imposing an upper limit to the length of the individual papers. Editors receive jointly 30 complimentary copies of their book. They are entitled to purchase further copies of their book at a reduced rate. As a rule no reprints of individual contributions can be supplied. No royalty is paid on Lecture Notes in Physics volumes. Commitment to publish is made by letter of interest rather than by signing a formal contract. Springer-Verlag secures the copyright for each volume.

The Production Process

The books are hardbound, and the publisher will select quality paper appropriate to the needs of the author(s). Publication time is about ten weeks. More than twenty years of experience guarantee authors the best possible service. To reach the goal of rapid publication at a low price the technique of photographic reproduction from a camera-ready manuscript was chosen. This process shifts the main responsibility for the technical quality considerably from the publisher to the authors. We therefore urge all authors and editors of proceedings to observe very carefully the essentials for the preparation of camera-ready manuscripts, which we will supply on request. This applies especially to the quality of figures and halftones submitted for publication. In addition, it might be useful to look at some of the volumes already published. As a special service, we offer free of charge LaTeX and TeX macro packages to format the text according to Springer-Verlag's quality requirements. We strongly recommend that you make use of this offer, since the result will be a book of considerably improved technical quality. To avoid mistakes and time-consuming correspondence during the production period the conference editors should request special instructions from the publisher well before the beginning of the conference. Manuscripts not meeting the technical standard of the series will have to be returned for improvement.

For further information please contact Springer-Verlag, Physics Editorial Department II, Tiergartenstrasse 17, D-69121 Heidelberg, Germany

John W. Clark   Manfred L. Ristig (Eds.)

Theory of Spin Lattices and Lattice Gauge Models

Proceedings of the 165th WE-Heraeus-Seminar
Held at Physikzentrum Bad Honnef, Germany,
14–16 October 1996

 Springer

Editors

John W. Clark
Washington University
St. Louis, MO 63130, USA

Manfred L. Ristig
Institut für Theoretische Physik
Universität zu Köln
D-50937 Köln, Germany

Cataloging-in-Publication Data applied for.

Die Deutsche Bibliothek - CIP-Einheitsaufnahme

Theory of spin lattices and lattice gauge models : proceedings of
the 165th WE-Heraeus-Seminar, held at Physikzentrum Bad Honnef,
Germany, 14 - 16 October 1996 / John W. Clark ; Manfred L. Ristig
(ed.).

(Lecture notes in physics ; Vol. 494)
ISBN 978-3-662-14113-7   ISBN 978-3-540-69211-9 (eBook)
DOI 10.1007/978-3-540-69211-9

ISSN 0075-8450
ISBN 978-3-662-14113-7

Typesetting: Camera-ready by the authors/editors
Cover design: *design & production* GmbH, Heidelberg
SPIN: 10550837   55/3144-543210 - Printed on acid-free paper

Preface

The contributions to this volume are based on invited talks presented at the 165th WE Heraeus Seminar, which was devoted to the topic "Theory of Spin Lattices and Lattice Gauge Models". The workshop was designed to bring together a number of prominent scientists from the international community working in selected areas of modern condensed-matter physics, both basic and applied. Emphasis was placed on three main aspects of the central topic, which mirror converging developments in several subfields of physics.

- Spin lattices are of basic interest in solid-state physics, especially for fundamental studies of magnetism. Spin arrangements on lattices interact in many different ways and exhibit a surprisingly rich array of phase transitions. Their study can provide a foundation for fruitful analysis of complex materials and for the creation of novel forms of matter.

- An exciting aspect of basic research on spin interactions is its applicability, notably through the invocation of duality relations, to lattice gauge problems in quantum field theory. In the same spirit of universality, recent research based on advanced many-body theories of strongly correlated systems has opened the way to unified theoretical approaches to apparently diverse many-body problems ranging from spin lattices to electronic systems of mobile carriers to condensed quantum liquids. Moreover, recent analyses of gauge fields and chiral meson fields have revealed a commonality with the physics of Josephson junction arrays in applied physics.

- Remarkable progress, experimental as well as theoretical, on these related classes of problems has brought us to a level of quantitative understanding that reaches far beyond the simple mean-field pictures that pervade the familiar literature. These advances, made possible by the realistic introduction of correlation effects, can and should be exploited to develop a deep microscopic description of the structural phase transitions in ferroelectrics, ferroelastics and other complex lattice systems that are now receiving concerted attention in crystallography and materials science. To a large extent, complex space-group symmetry analysis still dominates the latter fields. Exploration of phase transition mechanisms is, for the most part, conducted at a phenomenological level within Landau's approach to first- and second-order phase transitions.

In short, the accomplishments and the available expertise of scientists working on spin systems, lattice gauge models, and quantum liquids and solids has culminated in an extraordinary opportunity for rapid and efficient development of realistic strategies and algorithms for *ab initio* theoretical analysis of conventional and exotic condensed-matter systems. Such efforts can lend invaluable support to the systematic design and experimental synthesis of novel materials. The 165th WE Heraeus Seminar was intended to promote these ends through constructive interaction of researchers engaged in the relevant subfields.

The seminar provided a forum in which thirty-four invited participants could discuss the latest work on the highly interdisciplinary subject of lattice many-body systems. The setting of the meeting in the Honnefer Haus was most conducive to productive discussion, both formal and informal. Sixteen invited talks were delivered by internationally recognized experts. Eleven of the invited papers appear in this volume. Following the standard procedures of established physics journals, these papers have been refereed by independent researchers with specific knowledge of the problems being addressed. The oral presentations by K. Binder, R. F. Bishop, J. Carlson, P. de Forcrand, and J. Ihringer were of similarly high quality. However, written versions of these invited talks were not submitted, either because of time limitations for the preparation of manuscripts or because the results described have, to a large extent, been published elsewhere.

The meeting was made possible by the generosity of the WE Heraeus Stiftung. We wish to express our great appreciation to Dr. V. Schäfer, Dr. A. Bischoff, and Mrs. Jutta Hartmann from the Foundation for their exemplary efforts in the general organization of the seminar and for the congenial atmosphere that they fostered during the workshop itself. We also thank the participants along with all others who helped to make the seminar a success. Our appreciation also goes to Dr. Jin Woo Kim for his very efficient assistance in preparing this publication.

Washington University   J.W. Clark
Universität zu Köln   M.L. Ristig
May 1997

Contents

List of Participants

Baker, Stephen J.   MCBICSJB@afs.mcc.ac.uk
    Department of Physics, University of Manchester Institute of Science and
    Technology, Manchester, M60 1QD, United Kingdom,
Binder, Kurt   binder@chaplin.physik.uni-mainz.de
    Institut für Physik, Johannes-Gutenberg-Universität, Staudinger Weg 7, D-
    55099 Mainz, Germany,
Biró, Tamas S.   tsbiro@sunserv.kfki.hu
    Theory Division, MTA KFKI RMKI, Pf. 49, H-1525 Budapest, Hungary,
Bischoff, A.
    WE-Heraeus-Stiftung, Postfach 1553, D-63405 Hanau, Germany,
Bishop, Raymond F.   R.F.Bishop@UMIST.ac.uk
    Department of Physics, University of Manchester Institute of Science and
    Technology, Manchester, M60 1QD, United Kingdom,
Carlson, Joseph A.   carlson@qmc.lanl.gov
    Theoretical Division, Los Alamos National Laboratory, T5 MS B283, LANL,
    Los Alamos, NM 87505, USA,
Chin, Siu A.   chin@phys.tamu.edu
    Department of Physics, Texas A&M University, College Station, TX 77843,
    USA,
Clark, John W.   jwc@howdy.wustl.edu
    Department of Physics, Washington University, St. Louis, MO 63130, USA,
de Forcrand, Philippe   forcrand@scsc.ethz.ch
    SCSC, ETH-Zentrum, ETH Zürich, CH-8092 Zürich, Switzerland,
Dittmar, Susan   sdm@thp.uni-koeln.de
    Institut für Theoretische Physik, Universität zu Köln, Zülpicher Str. 77, D-
    50937 Köln, Germany,
Farnell, Damian J. J.   mccmmdf@afs.mcc.ac.uk
    Department of Physics, University of Manchester Institute of Science and
    Technology, Manchester, M60 1QD, United Kingdom,
Gernoth, Klaus A.   gernoth@thp.uni-koeln.de
    International School for Advanced Studies (SISSA), Via Beirut 2-4, I-31014
    Trieste, Italy, and Institut für Theoretische Physik, Universität zu Köln,
    Zülpicher Str. 77, D-50937 Köln, Germany,
Göhmann, Frank   frank.goehmann@theo.phy.uni-bayreuth.de
    Department of Physics, Faculty of Sciences, University of Tokyo, 7-3-1 Hongo,

Bunkyo-ku, Tokyo 113, Japan,

Gonsior, Bernhard
Eifelstr. 14, D-50677 Köln, Germany,

Hallberg, Karen   karen@mpipks-dresden.mpg.de
Max-Planck-Institut für Physik komplexer Systeme, Bayreuther Str. 40, Haus 16, D-01187 Dresden, Germany,

Haussühl, Siegfried   hauss@geocip.geo.uni-koeln.de
Institut für Kristallographie, Universität zu Köln, Zülpicher Str. 49b, D-50674 Köln, Germany,

Ihringer, Jörg P.   joerg.ihringer@uni-tuebingen.de
Institut für Kristallographie, Universität Tübingen, Charlottenstr. 33, D-72076 Tübingen, Germany,

Kim, Jin Woo   kim@thp.uni-koeln.de
Institut für Theoretische Physik, Universität zu Köln, Zülpicher Str. 77, D-50937 Köln, Germany,

Kröger, Helmut   hkroger@phy.ulaval.ca
Departement de Physique, Université Laval, Quebec, Quebec G1K 7P4, Canada,

Kürten, Karl E.   kuerten@acpx.exp.univie.ac.at
Institut für Experimentalphysik, Boltzmanngasse 5, A-1090 Wien, Austria,

Kusmartsev, Fedor V.   f.kusmartsev@lboro.ac.uk
Department of Physics, Loughborough University, Loughborough, Leis. LE11 3TU, United Kingdom,

Lindenau, Thomas   tl@thp.uni-koeln.de
Institut für Theoretische Physik, Universität zu Köln, Zülpicher Str. 77, D-50937 Köln, Germany,

Manousakis, Efstratios   stratos@ithaca.martech.fsu.edu
Department of Physics, 318 Keen Building, Florida State University, Tallahassee, FL 32306, USA,

March, Norman H.
Oxford University, England, 6 Northcroft Road, Egham, Surrey TW 20 0DU, United Kingdom,

Mavrommatis, Eirene   emavrom@atlas.uoa.gr
Physics Department, University od Athens, Panepistimioupolis, Ilissa, GR-15771 Athens, Greece,

Mertens, Franz G.   franz.mertens@theo.phy.uni-bayreuth.de
Lehrstuhl Theoretische Physik I, Universität Bayreuth D-95440 Bayreuth, Germany,

Oitmaa, Jaan   otja@newt.phys.unsw.edu.au
School of Physics, University of New South Wales, Sydney, NSW 2052, Australia,

Ristig, Manfred L.   ristig@thp.uni-koeln.de
Institut für Theoretische Physik, Universität zu Köln, Zülpicher Str. 77, D-50937 Köln, Germany,

XI

Schultka, N
Institut für Theoretische Physik, Technische Hochschule Aachen, D-52056 Aachen, Germany,

Serhan, Mohammed   `serhan@mpipks-dresden.mpg.de`
Max-Planck-Institut für Physik komplexer Systeme, Bayreuther Str. 40, Haus 16, D-01187 Dresden, Germany,

Shen, S
Max-Planck Institut für Physik komplexer Systeme, Bayreuther Str. 40, Haus 16, D-01187 Dresden, Germany,

van Dongen, Peter
Theoretische Physik III, Universität Augsburg, D-86135 Augsburg, Germany,

Wang, Wenguo   `ww@thp.uni-koeln.de`
Institut für Theoretische Physik, Universität zu Köln, Zülpicher Str. 77, D-50937 Köln, Germany,

Zheng, Weihong   `zwh@newt.phys.unsw.edu.au`
School of Physics, University of New South Wales, Sydney, NSW 2052, Australia,

Zheng, Bo   `zheng@pollux.physik.uni-siegen.de`
Fachbereich 7, Physik, Universität Siegen, D-57068 Siegen, Germany

Ising, Heisenberg and Hubbard Models in Relation to Insulating and Metallic Ferro- and Antiferro-magnets

N. H. March[1] and D. J. Klein[2]

[1] Oxford University, Oxford, England

[2] Department of Marine Sciences, Texas A&M University at Galveston, Galveston, TX 77553-1675, USA

Abstract. The Ising model in low dimensions is used for ferromagnets to relate internal energy and entropy to the magnetization. While this is done throughout the ferromagnetic phase, the low temperature predictions are compared with microscopic elementary excitations theory for both insulating and metallic ferromagnets. The model predictions are oversimplified. The spin $s = \frac{1}{2}$ Heisenberg model for an insulating antiferromagnet is then considered, starting from one dimension and building up a two-dimensional square lattice from lattice strips of variable width. Chemical approaches based an counting local spin-pairing patterns (or Kekulé structures) are brought into contact with recent work on ladders, with both even and odd numbers of legs, in the context of high-T_c cuprates. Finally, the Hubbard model and the closely related t-J model are discussed. For the former, simple rules, again based on a chemical approach, are proposed for predicting the spin properties of the ground states and comparison is made with existing computer studies. The related t-J model is briefly considered in relation to carriers moving through antiferromagnetic assemblies as in the high-T_c materials.

1 Introduction

The aim of this paper is to bring the predictions of three well-known models, Ising, Heisenberg and Hubbard, into direct contact with experiments, both *real* and using the computer, for insulating and metallic ferro- and antiferro-magnets.

The Ising model being the simplest, with exact solutions available in one- and two-dimensions (Reichl 1980), it is the natural starting point here. Its predictions will first be used for ferromagnets to relate (a) the deviation ΔE of the internal energy from its ground-state value $E(0)$, i.e., $\Delta E = E(T) - E(0)$ and (b) the entropy S, to $\Delta M = M(0) - M(T)$, with M the magnetization, for temperatures throughout the ferromagnetic phase, i.e., $0 \leq T \leq T_c$ with T_c the Curie temperature. At low temperatures, microscopic theory based on well-established elementary excitations is used to correctly relate ΔE and S to ΔM for (a) insulating and (b) metallic three-dimensional ferromagnets. The Ising model predictions are different as T tends to zero and the differences do not arise from dimensionality considerations.

The second area treated below invokes the Heisenberg model for an insulating antiferromagnet with spin $s = \frac{1}{2}$. Starting again from one dimension, the building up of a two-dimensional square lattice, realistically describing the Cu spins in CuO_2 planes in the parent insulators of high-T_c materials, is considered, starting from rectangular fragments of width W cut from the planar lattice, and following the early work of Živković et. al. (Živković 1989), (Živković, et. al. 1990) contact is made with very recent work on ladders, with both even and odd numbers of legs, again in the context of high-T_c solids (Dagotto, et. al. 1996). The low-lying excitations are qualitatively different for even and odd legs in the ladders, and the approach to the square-planar lattice limit is therefore irregular. While by judicious choices of steps, the irregular behavior can be dealt with, it may turn out that even leg extrapolation is more appropriate for doped materials while the odd leg extrapolation describes the insulating parent antiferromagnetic cuprates.

Pressing the chemical (valence-bond) point of view, some simple rules are then proposed for predicting the spin properties of the Hubbard model. Where it proves feasible, contact is established with the results of computer experiments. Finally, the related t-J model will be discussed in the context of carriers moving through antiferromagnetic assemblies and compared with Fermi liquid theory predictions relating the product of electrical resistivity and nuclear spin-lattice relaxation time directly to temperature T (Egorov and March 1994), (March 1996). This motivates a proposal to test critically the applicability of the t-J model to describe the normal state of high-T_c materials.

2 Ising model predictions and relation to low temperature properties of insulating and metallic ferromagnets

As Grout and March (Grout and March 1976) pointed out, the exact solution of the two-dimensional Ising model (Reichl 1980) can be used to plot the internal energy E versus the reduced magnetization $[M(0) - M(T)]/M(0)$ throughout the ferromagnetic phase. They noted the linearity of the plot near $T = 00$, and this feature is, in fact, independent of dimensionality and characteristic of Ising-like models.

2.1 Comparison with microscopic theory of low-lying excitations

Let us immediately bring the Ising prediction $\Delta E \sim \Delta M$ at low temperatures, with

$$\Delta E = E(T) - E(0) \quad : \Delta M = M(0) - M(T) \tag{1}$$

into contact with microscopic theory using well-established properties of low-lying excitations in (a) insulators and (b) metallic ferromagnets (three-dimensional: e.g., Ni in case(b)).

(a) Insulating ferromagnet

Bloch applied spin-wave theory (Callaway 1974) to calculate the low-lying excitations of a three-dimensional ferromagnetic insulator, with the low-temperature result for the magnetization (Callaway 1974),

$$\Delta M(T) = \text{const} \cdot T^{\frac{3}{2}} + \cdots \quad . \tag{2}$$

Similarly the specific heat $C_v \propto T^{3/2}$ (Callaway 1974) and hence

$$\Delta E(T) = \text{const} \cdot T^{\frac{5}{2}} + \cdots \quad . \tag{3}$$

It evidently follows from equations (2) and (3) that

$$\Delta E = \text{const} \cdot (\Delta M)^{\frac{5}{3}} \tag{4}$$

in contrast to the linear relationship of the Ising model.

(b) Metallic ferromagnet

While the so-called Stoner single-particle excitations contribute a term of $0(T^2)$ to $\Delta M(T)$ in Eq. (2), this latter equation is still valid at sufficiently low temperatures in metallic ferromagnets. But the electron liquid now at elevated temperatures has single-electron excitations within $k_B T$ of the Fermi energy ϵ_f and hence the classical equipartition of energy yielding $\Delta E_{class} \sim k_B T$ is modified to read

$$\Delta E(T) \sim (k_B T)(k_B T/\epsilon_f) \sim T^2 + \cdots \quad . \tag{5}$$

Combining therefore Eq. (5), to be contrasted with Eq. (3) for the insulator, with Eq. (2) yields for the metallic case

$$\Delta E \sim (\Delta M)^{\frac{4}{3}} \tag{6}$$

which is different from both the insulating ferromagnet results (4) and the Ising linear form.

Recently, March et.al. (March, et. al. 1996) have utilized the Ising model with an applied magnetic field in one dimension to plot the entropy against magnetization for spin $\frac{1}{2}$, and to relate to the free-spin result (Tuszynski 1991). We shall not go into details here, but merely remark that the entropy at low temperatures can again be obtained from the microscopic theory of elementary excitations and compared with the Ising model predictions.

3   Spin-1/2 Heisenberg antiferromagnet on square-planar lattice

Let us turn from the ferromagnet discussed above to the antiferromagnetically signed Heisenberg spin Hamiltonian. Anderson (Anderson 1987) proposed that this model had relevance to the understanding of the high T_c behavior of the ceramic perovskites, which are built from two-dimensional square-planar arrays of active Cu atoms, and subsequent work has amply confirmed his proposal (Živković, et. al. 1990).

This led Živković et. al. (Živković 1989) to investigate the ground-state of the Heisenberg model for fragments cut from the square-planar lattice, for infinite-length strips, and finally to extrapolate the strip width to the two-dimensional lattice itself. Interest in this approach has been rekindled by the discovery of the so-called ladder materials, e.g., the compound $(VO)_2P_2O_7$ (Grout and March 1976) which has two-leg ladders as does the cuprate $SrCu_2O_3$, or the three-leg ladder compound $Sr_2Cu_3O_5$ (see also Fig. 1 of Ref. (Dagotto, et. al. 1996)).

Two aspects will be focussed on below, the main points being exemplified using the *chemical* approach of Živković et. al. (Živković 1989): (i) the way *strips* can be chosen to avoid the irregular extrapolation to the square-planar lattice because of the different physics and chemistry apparent in ladders with even and odd numbered widths and (ii) related to (i), the qualitatively different excitations of even and odd widths of the ladder materials.

3.1   Neél-liquid and RVB-state descriptions

The understanding of the behavior of these ladders entails comparison of resonating valence-bond (RVB) and Neél-state descriptions. The RVB approach most simply invokes the use solely of local singlet-spin-pairing states, which for suitable (bipartite) lattices admitting A and B sublattices take the form (Klein, et. al. 1996)

$$|K\rangle \equiv \prod_i^A \{\alpha(i)\beta(Ki) - \beta(i)\alpha(Ki)\} \tag{7}$$

where $\alpha(i)$ and $\beta(j)$ denote spin-up and spin-down orientations for sites i and j, while the i-product is over all sites of the A sublattice, and K is a correspondence associating to each site i of the sublattice a corresponding nearest-neighbor site Ki of the B sublattice. Clearly there generally are many such correspondences K, each associating to a different pattern of coupling pairs of spins to a local singlet. As a first approximation the RVB ground-state may be viewed as a linear combination of these $|K\rangle$, most simply all with equal weights,

$$|\Psi_{RVB}\rangle = \sum_K |K\rangle \quad . \tag{8}$$

Comparison of the energy of this state with that of the Neél state might then be taken as an indicator of the relative adequacy of the RV Bond Neél-state descriptions. Pauling (Pauling 1949) identified the local spin-pairing states $|K\rangle$ as *Kekulé structures*: often referred to nowadays as *dimer coverings*. The simple *Ansatz* (Pauling and Wheland 1933) for $|\Psi_{RVB}\rangle$ is sometimes referred to as the Pauling-Wheland *Ansatz*, or alternatively the zero-order RVB *Ansatz*.

The expectation values for the Heisenberg model

$$H = J \sum_i^A \sum_j^i 2S_i S_j \tag{9}$$

are desired, J being the exchange parameter (which is positive for the cases of interest). Then the energies per site are

$$\mathcal{E}(\text{Neél}) = -\frac{cJ}{4} \tag{10}$$

and

$$\mathcal{E}(\text{RVB}) = -\frac{3J}{4} + \mathcal{E}_{res} \tag{11}$$

where c is the coordination number of the lattice L and $-3J/4$ is the expectation for any one of the (degenerate) $|K\rangle$, while $\mathcal{E}_{res}$ is the additional *resonance* stabilization due to interaction of the various $|K\rangle$. Evidently the greater the number of these K the greater the extent of *resonance*, so that $\mathcal{E}_{res} \leq 0$ should typically be more stabilizing for larger c, though in fact further details of the lattice structure should matter. Evidently (Klein, et. al. 1985) the RVB picture can be anticipated to be superior for $c \leq 3$ (which is just the realm for the carbon π-networks of organic chemistry), but for the square-planar lattice with $c = 4$ the Neél-state picture seems superior. Of course neither the simple RVB wavefunction nor the simple Neél-state wavefunctions are exact eigenstates – they are viewed just as zero-order *Ansätze*: when needed both could undergo further refinement. For the Neél-state description one might imagine the long-range magnetic order to be lost, though we still speak af a *Neél liquid*.

3.2   Ladders, lattice fragments and elementary excitations

The various ladder strips cut from the square-planar lattice would however have an intermediate value of coordination number $3 \leq c \leq 4$. And the doped square-planar lattice (relevant for high-temperature superconduction) with unpaired spins absent on some sites would also have an effective coordination number for local spin-pairing slightly less than $c = 4$.

Some further estimate of *resonance* can be obtained through the consideration of a long-range ordering inherent in the local spin-pairing patterns. This has previously been discussed (Klein, et al. 1986), at a general formal level, but is perhaps most easily understood via a few examples, say on a width-3 strip as in figure 1. There a typical dimer-covering pattern is displayed, and dashed lines are drawn so as to divide the strip unit cells. Notably the number of *dimers* (corresponding to local singlet-spin pairs) crossing the unit-cell boundaries alternates, between 1 and 2, and continues to do so no matter how far the dimer pattern is continued along the strip. That is, for this (or any) K there is a long-range ordering in the number of local spin-pairings crossing the boundaries of the unit cells. Of course there are different dimer coverings with different numbers of dimers crossing the unit cell boundaries – indeed the number Q_α of such crossings of a given unit-cell boundary α on a strip of width W may range from $Q_\alpha = 0$ to $Q_\alpha = W$. Kekulé structures (or dimer coverings, or spin-pairing patterns) with different values of Q_α correspond to different long-range orderings and so will not

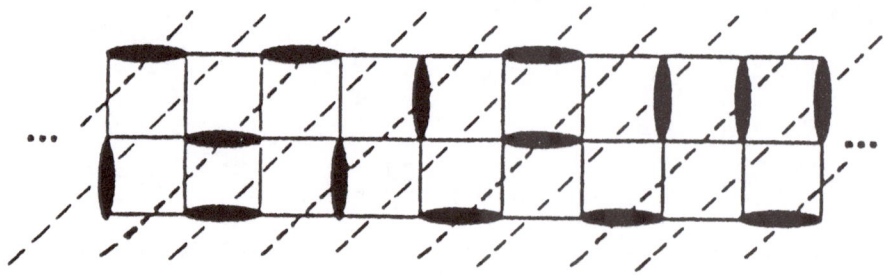

Fig. 1. A typical spin-pairing pattern on a width $W = 3$ ladder

admix, and those of the most numerous class (where there would be most *resonance*) constitute the best candidate for the overall ground-state. At this point it is evident that for the width $W = 3$ strip there are two degenerate classes: one with $Q_\alpha = 1$ yielding an alternation pattern for Q-values $1,2,1,2,\cdots$ starting from position α); and a second with $Q_\alpha = 2$ (yielding a corresponding pattern of Q-values $2,1,2,1,\cdots$). For width $W = 2$, the $Q_\alpha = 1$ class is non-degenerate and clearly gives many more dimer coverings than either of the other classes (with $Q_\alpha = 0$ or 2). Now generally a class with Q_α crossings at boundary α for a width W strip may be seen to give $Q_\beta = W - Q_\alpha$ crossings at the neiboring boundaries (because of the W sites per unit cell each is to be dimer-coupled to another site in either the cell to the right or to the left). But rather generally one might expect the number (per site) of dimer-covering patterns to be greater the more uniformity the *dimers* are distributed, and in particular, the more nearly Q_α and $Q_\beta = W - Q_\alpha$ are equal. Thus a greater degree of resonance is expected the smaller is $|Q_\alpha - Q_\beta|$, so that for the per site resonance energy we might expect

$$\mathcal{E}_{res} = -\mathcal{E}_{res}^o + C(\varDelta q)^2 \tag{12}$$

with $\mathcal{E}_{res}^o$ and C being positive constants encoding some other generic features of the general lattice, and where $\varDelta q$ is a *per-site* difference $(Q_\alpha - Q_\beta)/W$. Indeed such has been observed (Živković 1989). ¿From this we then suspect greater resonance stabilization for the $W = 2$ strip than for the $W = 3$ strip. Indeed all the even-width strips have a class with $Q_\alpha - Q_\beta = 0$ while for odd-width strips the minimum for $|Q_\alpha - Q_\beta|/W$ is $1/W$.

Next note that the long-range order is *spoiled* through the introduction of excitations consisting of locally unpaired spins. See, e.g., figure 2 for a $W = 2$ strip, where an unpaired spin (on the *asterisked* site) shifts the Q_α values from a sequence of $\cdots, 1,1,1,1,1$ on the left to a sequence $0,2,0,2,0,2,\cdots$ on the right.

In general the long-range ordering Q is shifted by +1 or -1 upon introduction of a single locally unpaired spin, on the A or B sublattice.

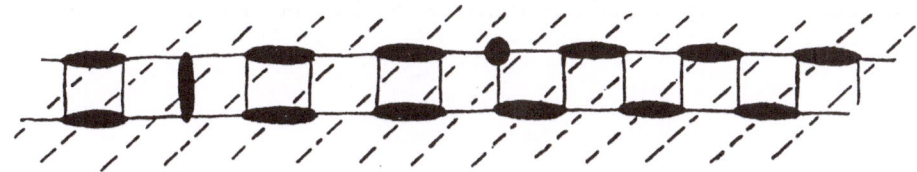

Fig. 2. A spin-pairing pattern with one locally unpaired spin, mediating the change from one type of spin-pairing pattern to another type.

In the case with a non-degenerate minimum energy class (as for $W = even$) this would result in the higher-energy long-range-ordered region, which could however be terminated through introduction of a second unpaired spin (on the opposite sublattice). That is, in such a non-degenerate case these spinon excitations would exhibit *confinement*, the attractive energy of interaction ever increasing (linearly) with increasing separation. Thus for width $W = 2$ the excitations should only be bound pairs of such spinons, whereas for width $W = 3$ there are two degenerate Q-phases, whence lower-energy single-spinon excitations should occur. Of course for sufficiently wide strips the oscillation between even and odd widths should be quenched.

In conclusion it is seen that especially for smaller-width chains there are two qualitatively distinct circumstances describing the indicated strips cut from the square-planar lattice. This is shown in Table 1. As the width become larger the

Table 1. Magnetic ordering and elementary excitations for ladders with even and odd number of legs

Width W	Long-range oder Δq	Ground-state	Excitations
Even	0	RVB	Higher-energy bound spinon pairs
Odd	$\frac{1}{W}$	Neél liquid	Lower-energy free spinons

difference between the strips diminishes with the Neél view eventually becoming more reasonable. The argument here is fairly qualitative, much of it is supported by quantitative computation (Živković 1989), (Dagotto, et. al. 1996) – and indeed much work (Klein, et al. 1997) has also been done for different types of strips cut from the hexagonal (honeycomb) lattice. For different types of strips cut from the square-planar lattice there seems to be little computation so far, though the same types of qualitative arguments can be applied. E.e., for the two types of strips of figure 3, such argument reveals that there should here be less difference between even- and odd-width chains. For the first type of strip both

Fig. 3. Two types of strips where the dramatic oscillations between even- and odd-widths should be quenched.

widths should be better described by a Neél-state description (because there are very few dimer coverings, as in turn can be seen from the extreme non-uniformity of the *ordering* for cuts along the long axis of these strips). For the second type of strip both even- and odd-widths when smaller, should favor a resonating VB type of ground-state.

To conclude this section, we return to the distinction between insulating and metallic magnets already emphasized for ferromagnetic long-range order in section 2. It appears that extrapolation to the square lattice limit by the odd-width ladders (only) may be physically appropriate for the parent insulating cuprate with antiferromagnetic ordering. However, for the doped materials, the even-width ladder extrapolation may be physically appropriate to recover the two-dimensional square lattice limit.

4 Simple chemical rules proposed to predict spin states of Hubbard model

We shall give below a summary of very recent work (Klein, et al. 1997) in which chemical rules have been proposed to predict spin properties of the Hubbard model. Important background to this study is the work of Nagaoka (Nagaoka 1966). He considered (a) the extreme $U = \infty$ limit and (b) a single hole (or a single electron) in a half-filled Hubbard band. Though this latter band is certainly not a fully saturated ferromagnet in experimentally realized circumstances, Nagaoka proved the remarkable result that one hole in an infinite lattice, measured from half-filling, leads to a completely polarized, i.e., fully saturated ferromagnetic state. Nagaoka's arguments do not apply, however, to completely general lattices and a more complete classification of the configurations on which Nagaoka's result is rigorous has been presented by Tasaki for one hole. The results proposed by Klein et. al. (Klein, et al. 1997) allow the prediction of spin multiplicity (i.e., spin magnetization) in the ground-state of cluster in higher dimensions from the results for a cyclic chain containing one or more holes.

4.1 Statement of rules

The rules proposed in (Klein, et al. 1997) may be summarized as follows. Consider , in the cluster being considered, the smallest cyclic chain of sites which can be constructed. If there is one hole created in the total cluster, apply the cyclic chain rule given in (Klein, et al. 1997) and set out in Table 2 to that smallest cyclic with its defect. There N_e and N_s are the numbers of electrons and sites in a cycle.

Table 2. Spin Multiplicity Rule for the $U = \infty$ Limit

small rings	$t > 0$	$t < 0$
$N_e = N_s - 1$	$S = N_e/2,\ S = \min$	$S = \min,\ S = N_e/2$
$N_e = N_e + 1$	$S = \min,\ S = (N_e - 2)/2$	$S = (N_e - 2)/2,\ S = \min$

If the spin predicted for that smallest cycle with a single defect is maximal, than the hole is evidently causing mutual alignment of the spins on singly occupied sites. The proposed rules then predict that, for the cluster being treated, the total spin is maximal also, so that one simply takes half of the total number of electrons to find the total spin of the cluster. This prediction has been brought into contact with specific results obtained for the full Hubbard Hamiltonian for tetrahedral and octahedral clusters.

When the opposite situation obtains: i.e., the spin predicted for the cycle with a single hole is minimal, then that defect is leading to spin pairing of the singly occupied sites. In this case, the prediction for the total spin of the cluster is 1/2 for an odd number of electrons and zero for an even number.

4.2 Some examples, using numerical solutions of Hubbard model

Let us simply enumerate examples considered in (Klein, et. al. 1996) in terms of the Hubbard model (with an electron-hopping parameter t and an on-site electron repulsion U):

(i) Octahedron for $t < 0$: single-band Hubbard model

Consider the spin of the ground-state for the octahedron for $t/U \neq 0$. Numerical studies by Callaway et. al. (Callaway, 1987) show, for the case $t < 0$, that for a single hole deviating from the one electron per site, the total spin $S = 1/2$ for all t/U. The rules proposed in (Klein, et al. 1997) yield this value, as the smallest cycle with its defect is, in this particular example, a triangle with one hole. Callaway et. al. (Callaway, 1987) have considered also the case of two holes. They observed in their numerical study a transition from the atomic limit with spin $S = 1$ to $S = 0$, as the electron-electron repulsion passes through $t/U = 0.0065$. This example shows that Nagaoka's theorem is rather robust, applying at finite (and substantial) concentration of holes. This has been demonstrated in fact by Anderson et. al. (Anderson, Shastry and Hristopulos 1989) using a variational approach.

Callaway et. al. (Callaway, et al. 1990), again for the octahedron with $t < 0$, also studied a single electron added to a band which is initially half-filled. Their finding was a ground-state having spin $S = 5/2$ for $|t|/U < 0.089$. This is again in agreement with the proposed rules (Klein, et al. 1997).

(ii) Tetrahedron for $t < 0$

For the tetrahedral configuration studied by Falicov and Victora (Falicov and Victora 1984), the proposed rules single out one triangular face with a single defect (Klein, et al. 1997). If the defect is a hole, the rules predict a low spin configuration as the ground-state, which is a doublet in this case. If the defect is an added electron (as usual measured from a half-filled band) the prediction is that the ground-state has maximum multiplicity, which in this example is a quartet. Both of these predictions (Klein, et al. 1997) agree with the numerical studies made by Falicov and Victora (Falicov and Victora 1984).

To briefly summarize the content of this section, the work of Klein et al. (Klein, et al. 1997) sheds light on the spin magnetization in finite clusters in the strong coupling regime of the Hubbard model. However, it leaves open questions as to the way in which one should pass from finite clusters to the thermodynamic limit. In the clusters studied to date, even with one hole in the reference state with one electron per site, the completely saturated ferromagnetic state is never found. However, this is the known ground-state of some lattices (Klein, et. al. 1985) in the thermodynamic limit, as follow from Nagaoka's theorem (Nagaoka 1966).

5 Relevance of t-J model to normal state of high-T_c materials

The t-J model has recently been studied (Qiming Li, et. al. 1991) and noncanonical Fermi liquid behavior found in the antiferromagnetic case.

This work has prompted comparison of predictions of (canonical) Fermi liquid theory with the above, in the normal state. We summarize below the conclusions of Egorov and March (Egorov and March 1994), stemming from the related study of Kohno and Yamada (Kohno and Yamada, 1991) using conventional Fermi liquid theory of a doped antiferromagnet with delocalized electrons or holes.

As Egorov and March stress, one can relate two observable quantities, namely the normal state electrical resistivity R and the nuclear spin-relaxation time T_1 via the susceptibility $\chi(\mathbf{Q})$ at the antiferromagnetic wave vector $\mathbf{Q}$. The results of the Fermi liquid study are:

$$R = const \cdot T^2 \chi(\mathbf{Q}) \tag{13}$$

and

$$(TT_1)^{-1} = const \cdot \chi(\mathbf{Q}) \quad . \tag{14}$$

At the time of the work of Ref.(Egorov and March 1994), measurements of R and T existed on the same underdoped cuprate, but no results were available for the susceptibility. Therefore, Egorov and March eliminated $\chi(\mathbf{Q})$ between equations (13) and (14) to yield

$$RT_1 = const \cdot T \tag{15}$$

A plot of RT_1 vs T for the above underdoped cuprate confirmed experimentally the normal-state prediction (15) at temperatures well above the superconductivity transition temperature. However, the RT_1 vs T plot at lower temperatures T still greater than T_c exhibited a minimum which was interpreted (March, et. al. 1996) as the formation of 2e-Bosons, following the pioneering work of Nozieres and Schmitt-Rink (Nozieres and Schmitt-Rink 1986). Be that as it may, there appears in this underdoped cuprate a clear demonstration of a substantial range of applicability of conventional Fermi liquid theory for carriers flowing through an antiferromagnetic assembly.

Therefore, a critical test seems possible as to whether the t-J model is a valid quantitative tool in normal-state theory of high T_c cuprates. Very specifically, does the non-canonical Fermi liquid emerging from the t-J model applied to underdoped high T_c material also lead back to the relation (15), predicted by conventional Fermi liquid theory applied to carriers flowing through an antiferromagnetic assembly, the relation (15) being supported by experiment.

6 Summary

The models considered here: Ising, Heisenberg, Hubbard, and a variant of the latter, the so-called t-J model, have been brought into contact with both real and computer experiments and, in a more limited sense, also with the microscopic theory of elementary excitations near absolute zero. Stress has been placed on comparing ferro- and antiferro-magnetic behavior in insulators with that in metals.

Fore low temperature properties of ferromagnets, all approaches relate the deviations $\Delta E(T)$ and $\Delta M(T)$ from ground-state internal energy and magnetization respectively by power-law behaviors $\Delta E \sim (\Delta M)^{1+p}$. But $p = 0$ for Ising models, whereas for three-dimensional insulating ferromagnets $p = 2/3$ and for metallic magnets $p = 1/3$.

Turning to antiferromagnets, the Heisenberg $s = 1/2$ model with a square-planar lattice is well suited to describe the Cu spins in the CuO_2 layers of high-T_c parent insulators. However, surprises come up when one attempts to build the square planar lattice from infinite strips, as in the early work of Živković et al.(Živković, et. al. 1990). Ladder compounds are now known with both even and odd numbers of legs, examples being cited above. Even leg ladders exhibit a spin gap Δ for a finite number of rings, but $\Delta \to 0$ as the *width* of the ladder tends to infinity, whereas, odd-leg ladders have zero spin gap. Extrapolation is therefore irregular to the two-dimensional square lattice limit. Regularization in approaching this limit can be improved, as pointed out above, by careful choice of strip geometry.

Rules, based on largely chemical arguments, are proposed which are predictive for the spins of the ground-states of clusters described by the Hubbard model. Where possible, the predictions are brought into contact with computer experiments and, at the time of writing, complete agreement exists. Finally, the t-J model, a variant of the Hubbard Hamiltonian, is discussed in relation to high-T_c materials and a test of this model as a valid description of the normal state of such materials is proposed.

As to directions for further work, we want to draw attention to the possible relation between Heisenberg and Ising models, and more generally between other quantum models (such as the t-J model), plus Ising-like (also termed lattice gas) models. This might build on the work of Garcia-Bach and Klein (Garcia-Bach and Klein, 1996) who write a rather general wave function *Ansatz*. Then, with Ψ the ground-state wave function, $\langle \Psi | \Psi \rangle$ becomes a measure of the partition function $\mathrm{tr}(\exp(-\beta H_{Ising}))$. Here β serves as a parameter, though more generally additional parameters may be introduced in the Ising-like model. The interested reader should consult the paper by Garcia-Bach and Klein which is a first step in the new direction proposed.

Acknowledgment

One of us (NHM) wishes to acknowledge partial financial support from the Office of Naval Research for work on matter in high magnetic fields. He is most grateful to Dr. P. Schmidt for much discussion and support on this area.

References

Anderson P. W., Science **235**, 1196 (1987).

Anderson P. W., Shastry B. S., and Hristopulos D., Phys. Rev. **B 40**, 8939 (1989).

Callaway J., *Quantum Theory of the Solid State* (Academic, New York, 1974) p.136.

Callaway J., Chen D. P., Kanhere D. G., and Quiming Li, Phys. Rev. **42**,465 (1990).

Callaway J., Chen D. P., and Tang R., Phys. Rev. **B 35**, 3705 (1987).

Dagotto E. and Rice T. M., Science **271**, 618 (1996).

Egorov S. A. and March N. H., Phys. Chem. Liquids, **27**, 195 (1994).

Falicov L. M. and Victora R. H., Phys. Rev. **B 30**, 1695 (1984)

Garcia-Bach M. A. and Klein D. J., J. Phys. A: Math. Gen. **29**, 103 (1996).

Grout P. J. and March N. H., Phys. Rev. **B 14**, 4027 (1976).

Klein D. J., Schmalz T. G., Zhu H. Y., and March N. H., Paper presented at *60-years of Antiferromagnetism*, College Station, Texas, 1996.

Klein D. J., Schmalz T. G., Hite G. E., and Seitz W. A., Chem. Phys. Lett. **120**, 367 (1985).

Klein D. J., Alexander S. A., Seitz W. A., Schmalz T. G., and Hite G. E., Theor. Chem. Acta **69**, 393 (1986).

Klein D. J., Alexander S. A., and March N. H., to appear, 1997.

Kohno H. and Yamada K., Prog. Theor. Phys. **85**, 13 (1991).

March N. H., Phys. Chem. Liquids, in press.

March N. H., Nip A. M. L., Tuszynski J. A., Int. J. Quantum Chemistry, Sanibel Symposium, 1997, in press.

March N. H., Pucci R., and Egorov S. A., Phys. Chem. Liquids, 1996.

Nagaoka Y., Phys. Rev. **147**,392 (1966).

Nozieres P. and Schmitt-Rink S., J. Low Temp. Phys. **59**,195 (1986).

Pauling L., *The Nature of the Chemical Bond* (Cornell University press, Ithaca, 1949); J. Chem. Phys. **1**, 280 (1933).

Pauling L. and Wheland G. W., J. Chem. Phys. **1**, 362 (1933).

Qiming Li, Callaway J., and Lun Tan, Phys. Rev. **B 44**, 10256 (1991).

Reichl L. E., *A Modern Course in Statistical Physics* (University of Texas Press, Austin, 1980).

Tuszynski J. A. and Wierzbicki W., Am. J. Phys. **59**, 555 (1991).

Živković T. P., J. Mol. Structure **185**, 169 (1989).

Živković T. P., Sandleback B. L., Schmalz T. G., and Klein D. J., Phys. Rev. **B 41**, 2249 (1990).

Studies of Lattice Spin Systems Using Series Expansions

J. Oitmaa, Zheng Weihong, and C.J. Hamer

School of Physics,
The University of New South Wales,
Sydney, NSW 2052, Australia.

Abstract. Efficient cluster expansion techniques have been developed for quantum Hamiltonian lattice models. Applications to antiferromagnetic Heisenberg systems, based on expansion about the Ising limit, yield accurate results for both ground state properties and excitation spectra. Recent work on novel systems, including spin ladders and frustrated two-dimensional systems, will be described.

1 Introduction

Lattice spin systems, typified by the Heisenberg model and its variants and generalizations, provide the essential theoretical underpinning of our understanding of magnetism in insulating solids. At the same time such models yield insights into generic aspects of strongly interacting many-body systems. In recent years attention has been focussed on low dimensional systems: spin chains, spin "ladders", and 2-d systems (for overviews see Affleck (1989), Dagotto and Rice (1996), and Manousakis (1991)). In particular the discovery of superconductivity in the cuprates has led to much new work in 2-d antiferromagnetism.

In the absence of exact solutions, which are only available in special cases, approximate analytic and numerical methods provide our only information. As the systems are complex, and sometimes show quite subtle effects, a combination of complementary approaches is needed. The method of series expansions, essentially the derivation of rather long perturbation series followed by an extrapolation or summation, is one powerful method of this sort. In some cases, as will be discussed below, this approach is able to yield results of comparable or superior accuracy to other numerical methods.

The method we have employed in our work is a "linked-cluster" approach, in which only configurations based on connected configurations contribute. Such methods are ubiquitous in many-body theory but in the present context originates from work of Nickel (1980) and Marland (1981). These ideas were developed for lattice gauge models by Irving and Hamer (1984) and further developed and applied to both spin and gauge models by our group at UNSW (e.g. He, Hamer and Oitmaa (1990), Zheng, Oitmaa and Hamer (1991),Hamer, Oitmaa and Zheng (1994), and Oitmaa, Hamer and Zheng (1994)). Essentially the same approach was rediscovered by Gelfand, Huse and Singh (e.g. Singh (1989), Gelfand, Singh and Huse (1989), Gelfand, Singh and Huse (1990), and Singh

(1993)) who have also applied it to many spin problems. This formalism most naturally applies to $T = 0$ ground state properties, where the non-contribution of disconnected clusters can be easily proven. Nickel also proposed a formalism for deriving series for excited states and hence energy gaps which involves, however, also disconnected clusters and thus introduces additional complexity into the procedure. Very recently Gelfand (1996) has shown how a linked-cluster expansion can be developed also for energy gaps and can be used to compute complete dispersion relations for excitations.

This article is not intended to be a comprehensive review of all the work which has been carried out by our group and others. Rather we want to outline the basis of the method, briefly outline some of the technical details so that the reader can get a flavour for what is involved, and outline applications of the method to some interesting systems which we have studied recently.

2 The formalism. Ising expansions for Heisenberg systems

We consider some, as yet undefined, lattice model for which the Hamiltonian can be written

$$H = H_0 + xV \tag{1}$$

where the unperturbed system, described by H_0, can be solved exactly, V represents the perturbation and x is a perturbation parameter. It can be shown that the ground state energy of H can be written as a sum of contributions from connected clusters G (a set of lattice sites which are "connected" via terms in the Hamiltonian)

$$E(x) = \sum_{\{G\}} C_G \mathcal{E}_G(x) \tag{2}$$

where C_G is the number of embeddings of G in the (periodic) lattice of N sites, and $\mathcal{E}_G(x)$ is referred to as the "cumulant energy" of cluster G.

The same formula can be used for a finite cluster G

$$E_G(x) = \sum_{\{G'\}} C_{G'/G} \mathcal{E}_{G'}(x) \tag{3}$$

where $C_{G'/G}$ is the number of embeddings of cluster G' in G. This allows the cumulant energies to be obtained iteratively, starting from the smallest cluster and working upwards. This is illustrated for, a few cases, in Fig. 1.

1   2   3   4   5   6   7   8

$$E_1 = \mathcal{E}_1 \qquad\qquad\qquad \mathcal{E}_1 = E_1$$
$$E_2 = \mathcal{E}_2 + 2\mathcal{E}_1 \qquad\qquad \mathcal{E}_2 = E_2 - 2\mathcal{E}_1$$
$$E_3 = \mathcal{E}_3 + 3\mathcal{E}_1 + 2\mathcal{E}_2 \qquad \mathcal{E}_3 = E_3 - 3\mathcal{E}_1 - 2\mathcal{E}_2$$

Fig. 1. The first few clusters and computation of cumulant energies.

There are two distinct parts of the calculation. A dictionary of clusters needs to be generated and the embedding constants computed. Such data are required in many areas of theoretical physics and efficient computer algorithms and codes exist (Martin (1974), Rapaport (1987)). These are problems of exponential complexity, in the algorithmic sense, and place limitations on how far it is feasible to go. For example, the number of distinct connected clusters with 10 sites is already 11,716,571. Similarly the embedding constants can become enormous and, if needing to be counted directly, can require lengthy computation. Nevertheless current programs can easily generate in excess of 10^6 graphs/sec. of CPU time and counts of 10^6 per CPU second are achievable.

It is also necessary to have an efficient procedure for computing the cluster ground state energies $E_G(x)$, as power series in x to some order. This is done using standard Rayleigh-Schrödinger perturbation theory, as explained in some detail by Hornby and Barber (1985).

This data can then be combined to obtain a perturbation series for $E(x)$ to some high order, typically $10-20$. Other ground state properties can be obtained in a similar way, by including appropriate fields in H.

There are two formalisms available for the calculation of excited state. The original one, proposed by Nickel and improved by Irving and Hamer (1984), involves both connected and disconnected clusters. Gelfand (1996) has recently developed a linked-cluster expansion to compute the complete dispersion relation for excitations. The key point for this approach is to construct an effective Hamiltonian within the manifold of degenerate excited states of the unperturbed Hamiltonian on a cluster. In general there will be L such states, consisting of single spin flips from the unperturbed ground state. The matrix elements of H^{eff}, within this manifold can be expanded perturbatively in powers of x

$$H^{\text{eff}}(l',l) = \sum_{k=0}^{\infty} x^k H_k^{\text{eff}}(l',l) \tag{4}$$

with the $H_k^{\text{eff}}(l',l)$ generated recursively through

$$H_k^{\text{eff}}(n,l) = \langle n|V|\psi_{k-1}^{(l)}\rangle \qquad n,l \leq L$$

$$\langle n|\psi_k^{(l)}\rangle = \frac{1}{E_0 - E_n}\left(\langle n|V|\psi_{k-1}^{(l)}\rangle - \sum_{k'=0}^{k-1}\sum_{l'=1}^{L}\langle n|\psi_{k'}^{(l')}\rangle H_{k-k'}^{\text{eff}}(l',l) \right) \tag{5}$$

with E_0, E_n being the unperturbed energies of the degenerate states and other states respectively. The derivation of these results is given by Gelfand (1996). To obtain H^{eff} to order k one needs the wavefunction ψ to order $k - 1$. This is less efficient than in the Nickel formalism, where knowledge of ψ to order k yields the energy to order $2k+1$. However it has the advantage that disconnected clusters do not contribute. Having obtained the matrix elements of H^{eff} in site representation, it is possible to obtain the complete dispersion relation directly by Fourier transformation.

To make these remarks more specific we consider the Heisenberg antiferromagnet with spin anisotropy,

$$H = \sum_{\langle ij \rangle} S_i^z S_j^z + \frac{x}{2} \sum_{\langle ij \rangle} (S_i^+ S_j^- + S_i^- S_j^+) \tag{6}$$

where the sums are over nearest neighbour pairs on a regular lattice, and the S_i are quantum spin-S operators. For $x = 1$ we recover the fully isotropic antiferromagnet, while $x = 0$ is the Ising model, which is trivial at $T = 0$. This choice of breakup corresponds to an "Ising expansion", and is the main subject of this article. Other choices for the unperturbed Hamiltonian can be made. For example, one could decompose the Hamiltonian into a part consisting of non-interacting dimers, which can be solved exactly and a perturbation consisting of the residual interactions between dimers. This is referred to as a "dimer expansion" (Gelfand, Singh and Huse (1989)). Similarly one could generate a "plaquette expansion".

We have derived series to x^{14} for the ground state energy, the magnetization and the parallel susceptibility and to order x^{10} for the energy gap for the $s = \frac{1}{2}$ antiferromagnet on the 2-dimensional square lattice (Zheng, Oitmaa and Hamer (1991)). This involves 11,131 connected clusters with up to 14 sites. The series can be accurately represented by Padé approximants for $x < 1$ and thus evaluated for any choice of anisotropy x. The isotropic point $x = 1$ is a singular point in the function and analysis at or near $x = 1$ is more difficult. We have used a transformation proposed by Huse (1988) to a new variable

$$\delta = 1 - (1 - x^2)^{1/2} \tag{7}$$

In Figure 2 we show the magnetization versus δ, obtained from our series. The magnetization at the isotropic point is estimated to be 0.307 ± 0.001, a value consistent with other methods. The most accurate Monte Carlo calculations (Runge (1992)) give 0.3075 ± 0.0025.

3 Recent applications to novel low-dimensional antiferromagnets

In this section we will describe some of the recent work of our group on three interesting systems of current interest.

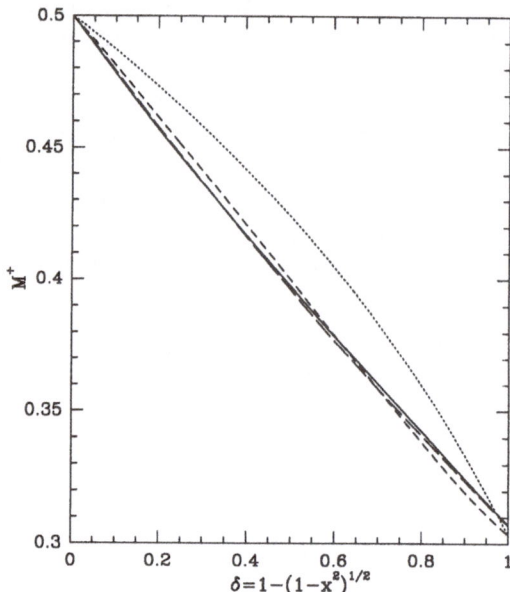

Fig. 2. Graph of the staggered magnetization M^+ against $\delta = 1 - (1 - x^2)^{1/2}$ for the spin-$\frac{1}{2}$ Heisenberg antiferromagnet on the square lattice. The four curves shown are the series estimate by Zheng, Oitmaa and Hamer (1991), and the first, second and third-order spin-wave predictions, corresponding to solid, dot, short dashed and long dashed lines respectively.

3.1 The $J_1 - J_2$ antiferromagnet

We consider the 2-dimensional spin-$\frac{1}{2}$ antiferromagnet with additional 2nd neighbour antiferromagnetic interactions of strength J_2. This interaction introduces frustration and competition between different types of ordering. Increasing J_2 will destabilize the Néel order at some critical value y_{1c} of $y = J_2/J_1$. For large J_2/J_1, on the other hand, the system will order in the "collinear phase" with alternating rows (or columns) of spins up and down, again with reduction of complete order by quantum fluctuations. As J_2/J_1 is reduced this phase will become unstable at some ratio y_{2c}.

In the Ising limit the two phases are stable for $y < 0.5$ and $y > 0.5$ respectively, i.e. $y_{1c} = y_{2c} = 0.5$, with no intermediate phase. A large number of studies of the Heisenberg case have suggested some intermediate phase, i.e. $y_{2c} \neq y_{1c}$. However the nature of the intermediate phase and the values of the critical points are far from clear. It has been suggested that the intermediate phase is a "spin liquid".

Using the Ising expansion method, as outlined in the previous section, we have obtained quite convincing evidence (Oitmaa and Zheng (1996)) that $y_{1c} \simeq 0.4$ and $y_{2c} \simeq 0.6$. This supports the existence of the intermediate phase and

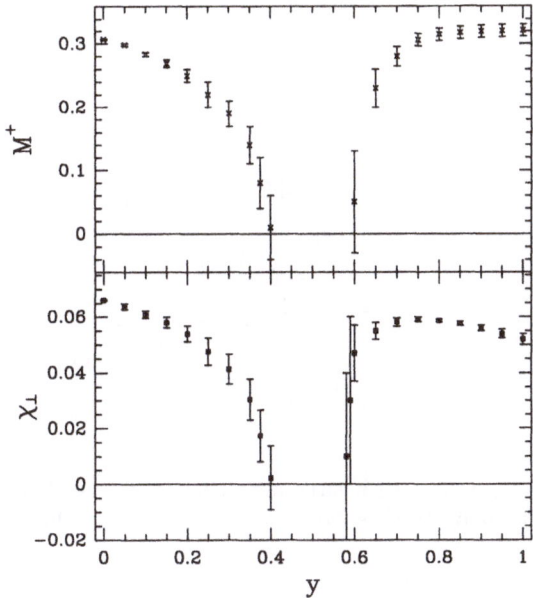

Fig. 3. The staggered/collinear magnetization M^+ and the perpendicular susceptibility $\chi_\perp$ for isotropic $J_1 - J_2$ model as function of coupling ratio $y = J_2/J_1$.

the estimates of y_{1c} and y_{2c} agree in fact very well with the best finite lattice diagonization results (Schultz and Ziman (1992)) on a 6×6 lattice.

Figure 3 shows our estimates for the appropriate order parameter versus coupling ratio y. This is obtained from Padé approximant analysis of the corresponding series in spin anisotropy x, for fixed y, evaluated as $x \to 1$. Even though the error bars are large, one can clearly see the vanishing of the order parameter as $y \to y_{1c}$ and $y \to y_{2c}$. We have also computed the 2nd and 3rd neighbour correlations and these also become small or zero at the phase boundaries, suggesting that the intermediate phase $0.4 \leq y \leq 0.6$ has only short range (essentially nearest neighbour) correlations.

3.2 The CAVO lattice

The material CaV_4O_9 is a highly anisotropic $S = \frac{1}{2}$ antiferromagnet. The magnetic layers can be represented as a square lattice with one-fifth of the spins removed, as in Fig. 4. The material exhibits a spin gap and it is of some interest whether this can be understood in terms of the spin degrees of freedom alone or whether spin-lattice coupling is required.

The general spin Hamiltonian for CaV_4O_9 can be written as,

$$\mathcal{H} = J_1 \sum_{(i,j)} \mathbf{S}_i \cdot \mathbf{S}_j + J_1' \sum_{(i,k)} \mathbf{S}_i \cdot \mathbf{S}_k + J_2 \sum_{(i,l)} \mathbf{S}_i \cdot \mathbf{S}_l + J_2' \sum_{(i,m)} \mathbf{S}_i \cdot \mathbf{S}_m, \quad (8)$$

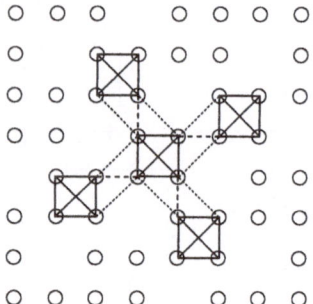

Fig. 4. The CAVO lattice, with sites indicated by circles. The couplings J_1, J_1', J_2, and J_2' are indicated by thick solid, thick dashed, thin solid, and thin dashed lines, respectively.

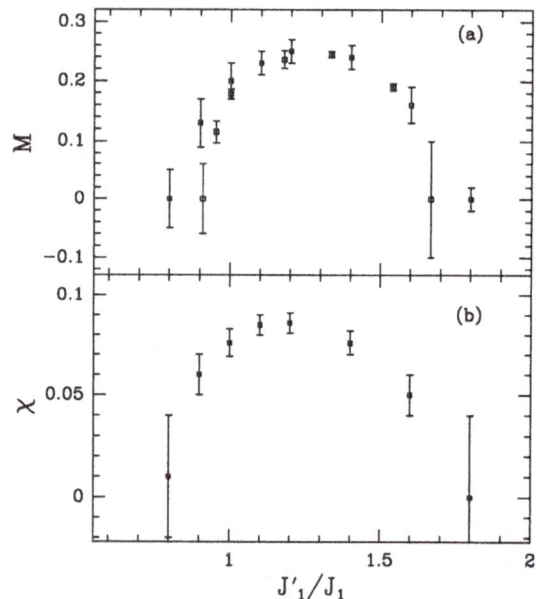

Fig. 5. Sublattice magnetization M and the uniform perpendicular susceptibility χ (at $T = 0$) *versus* J_1'/J_1 (with second-neighbor couplings $J_2 = J_2' = 0$) as estimated by Ising expansions (fulled symbols) and the quantum Monte Carlo results (open symbols) by Troyer, Kontani and Ueda (1996).

where the sums run over nearest-neighbor bonds within plaquettes (J_1), nearest-neighbor bonds between plaquettes (J_1'), second-neighbor bonds within plaquettes (J_2), and second-neighbor bonds between plaquettes (J_2').

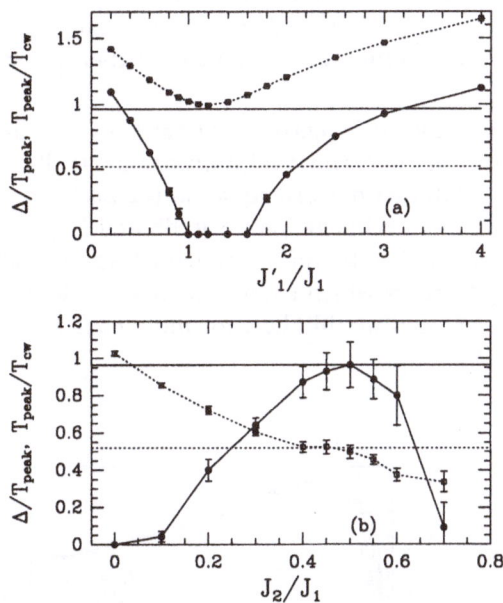

Fig. 6. Plots of ratios Δ/T_{peak} (shown as solid lines/full symbols) and T_{peak}/T_{cw} (shown as dash lines/open symbols) as a function of J_1'/J_1 (for systems with only nearest neighbor interactions) and J_2/J_1 (for systems with $J_2 = J_2'$ and $J_1' = J_1$). The horizontal lines indicate the experimental values.

The system has been studied by using four distinct convergent perturbation expansions to the maximum computationally feasible order (Gelfand, *et al.* (1996), and Zheng, *et al.* (1996)): High temperature expansions are used to calculate the temperature dependent susceptibility, Ising expansions are used to study phase boundaries and properties of the Néel-ordered phase, while plaquette and dimer expansions are used to calculate the ground-state properties and excitation spectra of magnetically disordered phases. To determine the exchange coupling for CaV_4O_9, the temperature dependence of the susceptibility and the spin gap are compared with the available experimental data. These expansions have been carried out for the following two sets of parameters:

i) nearest-neighbor interactions only ($J_2 = J_2' = 0$) for arbitrary $\lambda = J_1'/J_1$. The most precise and reliable results have been obtained for this case. Based on exponent-biased analyses of the high-order dimer and plaquette expansions for the triplet gap, we estimate the critical point dividing the plaquette phase at small λ from the Néel phase at intermediate λ to be $\lambda_{p-N} = 0.920(5)$, and the critical point dividing the Néel phase from the dimer phase at large λ to be: $1/\lambda_{N-d} = 0.60(1)$. The Ising expansions allow for estimates of the sublattice magnetization and the uniform perpendicular susceptibility $\chi_\perp^0$ within the Néel-ordered phase, as shown in Fig. 5. A Comparison of the theoretical results for the Curie-Weiss constant T_{CW}, the peak temperature T_{peak} and the gap Δ with the experimental data on CaV_4O_9 reveal that no exchange coupling for the model with nearest-neighbor couplings only can fit the experimental data well (see Fig. 6a).

ii) equal intra- and inter-plaquette couplings ($J_1 = J_1'$, $J_2 = J_2'$) for arbitrary J_2/J_1. For this case, the experimental data are generally consistent with that the second-neighbor interactions are equal to (or perhaps somewhat less than) half the nearest neighbor exchange, (see Fig. 6b and Fig. 7), but the fits are not entirely satisfactory. The low-temperature triplet and singlet excitation spectra have many striking features which should be observable in neutron and Raman scattering experiments and would allow for more confident estimation of model parameters.

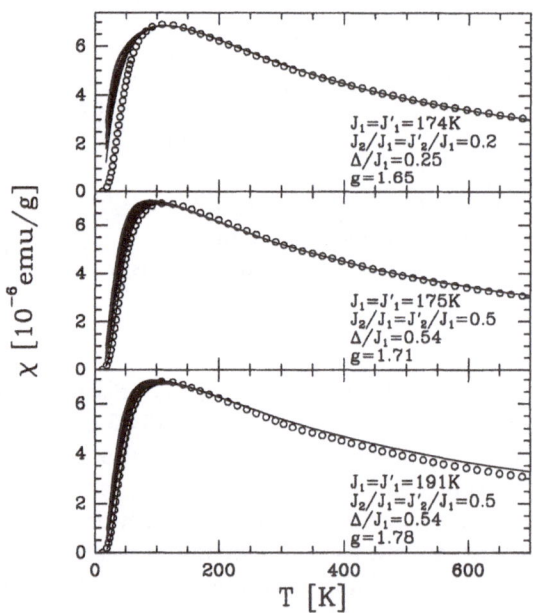

Fig. 7. Comparison of the calculated susceptibility for several Heisenberg models with parameters indicated in each panel (the solid lines representing various approximants), with the experimental data of Taniguchi *et al.* (1995) for CaV_4O_9.

3.3 Spin ladders

There has been considerable recent theoretical interest in coupled chain systems for a variety of reasons: Firstly the systems provide an interesting step from the relatively well understood one-dimensional behaviour towards two-dimensional systems (i.e. a dimensional crossover); a second reason for the interest is the possibility of realizing a lattice of weakly coupled ladders in compounds such as $(VO)_2P_2O_7$ and $SrCu_2O_3$.

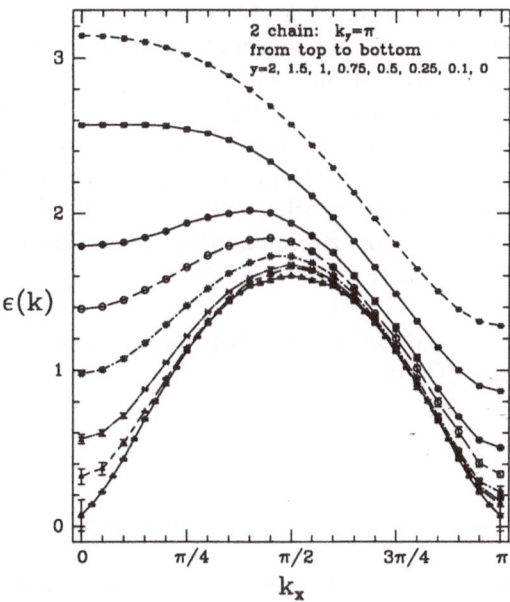

Fig. 8. The dispersions of the spin-triplet excited states of the 2-chain ladder with antiferromagnetic interchain coupling $y = J_\perp/J_{//}$ 2, 1.5, 1, 0.75, 0.5, 0.25, 0.1, 0 (shown in the figure from the top to the bottom respectively), for $k_y = \pi$, the results for $y \leq 1$ are from the dimer expansion, and the results for $y > 1$ are from the Ising expansion.

We have carried out extensive series studies of 2-chain and 3-chain ladder systems with both antiferromagnetic and ferromagnetic interchain coupling $J_\perp$, via Ising expansions and dimer expansions at $T = 0$, and also by high temperature series expansions (Oitmaa, Singh and Zheng (1996). Our results confirm the existence of a gap in the 2-chain system and delineate the phase diagram in the parameter space of Ising anisotropy and the parameter ratio $J_\perp/J_{//}$. The complete spin-wave excitation spectra, as shown in Fig. 8, are computed. For the 3-chain system we are unable to exclude the possibility of a small gap, although our results are consistent with a gapless spectrum. In addition, we develop a high-temperature series expansion for the uniform magnetic susceptibility and the specific heat for 2-chain and 3-chain systems with $J_{//} = J_{\perp}$; the susceptibility of 2-chain ladders is as expected for a system with a spin-gap while that of 3-chain ladders appear to remain finite in the zero-temperature limit suggesting an absence of a spin-gap.

4 Conclusion

This paper has given an overview of linked cluster perturbation methods used to study lattice spin and gauge models. Efficient computer codes have been developed for both the generation of cluster configurational data and for perturbation computations on clusters, and we have attempted to give the non-expert reader some flavour of the technical details. Our group at UNSW and the Davis/Colorado group have used these techniques over the last decade to elucidate the physics of many interesting systems. Some examples of recent studies have been given. In many cases the series expansion method can obtain results of equal or superior accuracy to other numerical methods, and this approach will continue to be used alone or in combination with other methods, in the studies of strongly interacting lattice systems.

This work forms part of a research project supported by a grant from the Australian Research Council.

References

Affleck, I. (1989): Quantum spin chains and the Haldane gap. J. Phys. Condens. Matter. **1**, 3047-3072.

Dagotto, E. and Rice, T.M. (1996): Surprises on the way from one- to two-dimensional quantum magnets - the ladder materials. Science **271**, 618-623.

Hamer, C.J., Oitmaa, J., and Zheng, W.H. (1994): Strong-coupling series for Abelian lattice gauge models in 3+1 dimensions. Phys. Rev. D **49**, 535-542.

He, H.X., Hamer, C.J. and Oitmaa, J. (1990): High-temperature series expansions for the (2+1)D Ising model. J. Phys. A **23**, 1775-1787.

Hornby, P.G. and Barber, M.N. (1985): Perturbation series for the mass gap of the (1+1)-dimensional O(2)-model. J. Phys. A **18**, 827-832.

Huse, D.A. (1988): Ground-state staggered magnetization of two-dimensional quantum Heisenberg antiferromangets, Phys. Rev. B **37**, 2380-2382.

Gelfand, M.P. (1996): Series expansions for excited states of quantum lattice models, Solid State Communications, **98**, 11-14.

Gelfand, M.P., Singh, R.R.P., and Huse, D.A. (1989): Zero-temperature ordering in two-dimensional frustrated quantum Heisenberg antiferromagnet. Phys. Rev. B **40**, 10801-10809.

Gelfand, M.P., Singh, R.R.P. and Huse, D.A. (1990): Perturbation expansions for quantum many-body systems. J. of Stat. Phys. **59**, 1093-1142.

Gelfand, M.P., et al. (1996): Convergent expansions for properties of the Heisenberg model for CaV_4O_9, Phys. Rev. Lett. **77**, 2794-2797.

Irving, A.C. and Hamer, C.J. (1984): Methods in Hamiltonian lattice field theory (II) Linked-cluster expansions. Nucl. Phys. B**230**, 361-384. Hamer, C.J. and Irving, A.C. (1984): Cluster expansions in the (2+1)D Ising model. J. Phys. A **17**, 1649-1664.

Manousakis, E. (1991): The spin-$\frac{1}{2}$ Heisenberg antiferromagnet on a square lattice and its application to the Cuprous Oxides. Rev. Mod. Phys. **63**, 1-62.

Marland, L.G., (1981): Series expansions for the zero-temperature transverse Ising model. J. Phys. A **14**, 2047-2057.

Martin, J.L. (1974): Phase Transition and Critical Phenomena, Vol. 3 (Domb, C. and Green, M.S., eds., Academic Press).

Nickel, B.G. (1980): unpublished.

Oitmaa, J. and Zheng, W.H. (1996): Series expansion for the $J_1 - J_2$ Heisenberg antiferromagnet on a square lattice, Phys. Rev. B **54**, 3022-3025.

Oitmaa, J., Hamer, C.J., and Zheng, W.H. (1994): Heisenberg antiferromagnet and the XY model at $T = 0$ in three dimensions. Phys. Rev. B **50**, 3877-3893.

Oitmaa, J., Singh, R.R.P., and Zheng, W.H. (1996): Quantum spin ladders at $T = 0$ and at high temperatures studied by series expansions. Phys. Rev. B **54**, 1009-1018.

Rapaport, D.C. (1987): Algorithm for Lattice Statistics. Computer Phys. Rep., **5**, 265-349.

Runge, K.J. (1992): Quantum Monte Carlo calculation of the long-range order in the Heisenberg antiferromagnet. Phys. Rev. B **45**, 7229-7236.

Schultz, H.J. and Ziman, T.A.L. (1992): Finite-size scaling for the two-dimensional frustrated quantum Heisenberg antiferromagnet. Europhys. Lett. **18**, 355-360.

Singh, R.R.P. (1989): Thermodynamic parameters of the $T = 0$, spin-$\frac{1}{2}$ square-lattice Heisenberg antiferromagnet, Phys. Rev. B **39**, 9760-9763.

Singh, R.R.P. (1993): Transverse-spin correlations and single-mode approximation for the square-lattice $S = \frac{1}{2}$ Heisenberg model. Phys. Rev. B **47**, 12337-12340.

Taniguchi, S. et al. (1995): Spin gap behavior of $S = \frac{1}{2}$ quasi-two-dimensional system CaV$_4$O$_9$, J. Phys. Soc. Jpn. **64**, 2758-2761.

Troyer, M., Kontani, H., and Ueda, K. (1996): Phase diagram of depleted Heisenberg model for CaV$_4$O$_9$, Phys. Rev. Lett. **76**, 3822-3825.

Zheng, W.H., et al. (1996): Heisenberg models for CaV$_4$O$_9$: expansions about high-temperature, plaquette, Ising, and dimer limits. submitted to Phys. Rev. B.

Zheng, W.H., Oitmaa, J. and Hamer, C.J. (1991): The square-lattice Heisenberg antiferromagnet at $T = 0$. Phys. Rev. B **43**, 8321-8330.

Application of Linked-Cluster Expansions to Quantum Hamiltonian Lattice Systems

Zheng Weihong, C.J. Hamer, and J. Oitmaa

School of Physics,
The University of New South Wales,
Sydney, NSW 2052, Australia.

Abstract. Comparisons are made between several different linked-cluster expansion methods, namely the linked-cluster perturbation series expansion, the t-expansion, the analytic Lanczos expansion, and the coupled-cluster expansion. They are considered from a technical point of view, and also as applied to the $S = \frac{1}{2}$ Heisenberg antiferromagnet on the square lattice and the compact U(1) lattice gauge model in 2+1 dimensions.

1   Introduction

Linked cluster expansions form an important part of our armoury of techniques for the investigation of quantum Hamiltonian systems in general, and lattice Hamiltonians at zero temperature in particular. There are several different methods of this sort on the market, including the standard perturbation series expansions (see Nickel (1980), Marland (1981), Irving and Hamer (1984), He, Hamer and Oitmaa (1990), and Gelfand, Singh and Huse (1990)), the "coupled cluster" expansion (see, for example, Coester and Kümmel (1960) and Bishop (1991)), the so-called "t-expansion" of Horn and Weinstein (1984), and the analytic Lanczos method (see Witte, Hollenberg and Zheng (1996)). There are quite close similarities between these techniques: they all calculate the change in some unperturbed initial state as the interaction is turned on in some way, and all rely on linked cluster theorems: for instance, the perturbation series for an extensive quantity such as the ground-state energy can be computed in terms of connected diagrams (or, the contributions from linked clusters of lattice sites) alone. It is important, therefore, to try and decide which technique requires the least computational effort, and which gives the most accurate results.

Our aim in this paper is to present some technical comparisons of these different approaches, and also compare the results of these methods when applied to the case of the $S = \frac{1}{2}$ XXZ Heisenberg antiferromagnet on the square lattice, and also the (2+1)-dimensional U(1) lattice gauge theory (LGT). We have recently computed some extended perturbation series and t-expansion series for these models. Zeng, Farnell and Bishop (1996), and Fang, Liu and Guo (1995) have also presented some coupled-cluster expansion results for these systems. Witte, Hollenberg and Zheng (1996) have carried out analytic Lanczos expansions for the Heisenberg antiferromagnet. By comparing the results of the different linked-cluster approaches, we can hope to decide which technique is the most "cost-efficient".

2 Comparison of Methods

2.1 The linked-cluster series expansion

Perturbation series expansions are of course a standard tool in every area of physics, and can generally be calculated most efficiently by linked cluster techniques (see, for example, Domb and Green (1974)). A linked cluster approach for quantum Hamiltonian lattice models was proposed by Nickel (1980), and further elaborated by Marland (1981) and Irving and Hamer (1984), and has been reviewed in some detail by He, Hamer and Oitmaa (1990). Gelfand, Singh and Huse (1990) independently evolved similar procedures. We present here a brief summary of the essential ideas.

Consider a lattice Hamiltonian of the form

$$H = H_0 + xV \tag{1}$$

where x is the expansion parameter, V is the perturbation, and H_0 is the unperturbed Hamiltonian. To obtain the series for some physical quantity such as the ground state energy E_0, we need the following information:

(i) a list of all clusters up to the order required. For the expansions considered here the clusters are grouped according to the number of sites or vertices. An efficient computer algorithm is used to generate all connected clusters;

(ii) the embedding constants C_α^N. These are the "strong" or low-temperature lattice constants, in standard terminology, and are enumerated by direct counting or algebraic reduction methods;

(iii) for each cluster α, a list of sub-clusters and corresponding embedding constants.

For a given cluster α, one can compute its ground state energy E_0^α by using standard Rayleigh-Schrödinger perturbation theory, and then get the "cumulant energy" of cluster α by subtracting the contribution from its subclusters:

$$\mathcal{E}_0^\alpha = E_0^\alpha - \sum_{\{\beta\}} C_\alpha^\beta \mathcal{E}_0^\beta$$

so that the ground state energy is:

$$E_0 = \sum_{\{\alpha\}} C_\alpha^N \mathcal{E}_0^\alpha \quad .$$

The calculation of the energy gap is a little more involved and we refer the reader to He, Hamer and Oitmaa (1990) and Gelfand (1996) for further details.

2.2 t-expansion

The t-expansion is a relatively recent technique, proposed originally by Horn and Weinstein (1984), and has been applied mainly to Hamiltonian lattice gauge theories. The method uses e^{-tH} to project any trial state $|\psi_0\rangle$ onto a new state,

$$|\psi_t\rangle \equiv \frac{e^{-tH/2}|\psi_0\rangle}{\langle\psi_0|e^{-tH}|\psi_0\rangle^{1/2}} \tag{2}$$

which converges to the true ground state $|\Psi_0\rangle$ of the Hamiltonian H as the "time" parameter t becomes large, provided that $\langle\psi_0|\Psi_0\rangle \neq 0$.

Therefore, the function

$$E(t) = \frac{\langle\psi_0|e^{-tH}H|\psi_0\rangle}{\langle\psi_0|e^{-tH}|\psi_0\rangle} \tag{3}$$

converges monotonically to the exact ground-state energy E_0 in the limit $t \to \infty$ with the following behaviour

$$E(t) \sim E_0 + ce^{-\Delta t}, \tag{4}$$

where Δ is the minimum energy gap in the ground state sector, and so we can compute the minimum energy gap in the ground state sector from $E(t)$ by the following relation:

$$-\frac{d^2 E(t)/dt^2}{dE(t)/dt} \to \Delta \quad \text{as} \quad t \to \infty \quad . \tag{5}$$

Expanding $E(t)$ as a power series in t, one finds

$$E(t) = \sum_{n=0}^{\infty}(-t)^n\frac{I_{n+1}}{n!} \tag{6}$$

where the coefficients I_{n+1} are the "connected moments" of the Hamiltonian and are defined recursively as:

$$I_{n+1} = \langle\psi_0|H^{n+1}|\psi_0\rangle - \sum_{p=0}^{n-1}\binom{n}{p}I_{p+1}\langle\psi_0|H^{n-p}|\psi_0\rangle \tag{7}$$

and the calculation of the expectation value $\langle\psi_0|H^n|\psi_0\rangle$ up to a certain order n will involve the wavefunction $H^i|\psi_0\rangle$ up to $i = n/2$. To get the wavefunction $H^i|\psi_0\rangle$, one only needs the knowledge of $H^{i-1}|\psi_0\rangle$, rather than the knowledge of $H^j|\psi_0\rangle$ up to order $j = i - 1$, as in the perturbation series expansion. So the t-expansion requires less computer memory compared to the perturbation series expansion.

To compute the connected moments, one needs the same list of clusters as in the linked-cluster perturbation series expansion. For a given cluster α and a chosen initial state $|\psi_0\rangle$, one can compute the connected moments by computing the expectation value $\langle H^n \rangle_0$, then get the "cumulant energy" of a cluster α by subtracting the contribution from its subclusters:

$$\mathcal{E}_\alpha(t) = E^\alpha(t) - \sum_{\{\beta\}} C_\alpha^\beta \mathcal{E}_\beta(t)$$

so that the ground state energy is:

$$E(t) = \sum_{\{\alpha\}} C_\alpha^N \mathcal{E}_\alpha(t) \quad .$$

As in the linked-cluster series expansion, the calculation needs to be very accurate in order to eliminate both round-off and overflow problems. This is usually done by using a symbolic manipulation package such as Maple. Here we employ multiple precision integer manipulations in Fortran.

2.3 Analytic Lanczos expansion

The analytic Laczos expansion was proposed by Hollenberg and Witte (1994), and the method is similar to that of the Lanczos method. If one starts from a trial state $|v_1\rangle$, and forms successive basis states $|v_n\rangle$ by

$$|v_n\rangle = \frac{1}{\beta_{n-1}}[(H - \alpha_{n-1})|v_{n-1}\rangle - \beta_{n-2}|v_{n-2}\rangle] \tag{8}$$

where α_n and β_{n+1} are defined by

$$\alpha_n = \langle v_n|H|v_n\rangle,$$

$$\beta_{n+1} = \langle v_{n+1}|H|v_n\rangle \tag{9}$$

then in this basis, one can prove that H is a tridiagonal matrix:

$$H \to T_n = \begin{bmatrix} \alpha_1 & \beta_1 & & & \\ \beta_1 & \alpha_2 & \beta_2 & & \\ & \beta_2 & \ddots & \ddots & \\ & & \ddots & \alpha_{n-1} & \beta_{n-1} \\ & & & \beta_{n-1} & \alpha_n \end{bmatrix} . \tag{10}$$

An expansion in the number of plaquettes (sites), N, on the lattice leads to the following forms for the Lanczos matrix elements $\alpha_n(N)$ and $\beta_n(N)$

$$\alpha_n(N) = c_1 + \frac{n-1}{N}\left[\frac{c_3}{c_2}\right]$$
$$+ \frac{(n-1)(n-2)}{N^2}\left[\frac{3c_3^3 - 4c_2c_3c_4 + c_2^2c_5}{4c_2^4}\right] + O(\frac{1}{N^3}) \tag{11}$$

$$\beta_n^2(N) = \frac{n}{N}c_2 + \frac{n(n-1)}{N^2}\left[\frac{c_2c_4 - c_3^2}{2c_2^2}\right] + \frac{n(n-1)(n-2)}{N^3}$$
$$\times \left[\frac{21c_2c_3^2c_4 - 12c_3^4 - 4c_2^2c_4^2 - 6c_2^2c_3c_5 + c_2^3c_6}{12c_2^5}\right] + O(\frac{1}{N^4}) \tag{12}$$

where c_n are the connected moments defined in t-expansion, so one can get some properties of the system by borrowing the connected moments from the t-expansion, making some truncation in α_n and β_n, then diagonalizing the tri-diagonal matrix T_n for finite N.

2.4 Coupled-cluster expansion

The coupled cluster expansion method has been extensively used in many-body theory, and has recently been introduced to lattice gauge theory by Bishop (see, for example, Bishop (1991)) and Guo *et al* (see, for example, Fang, Liu and Guo (1995)), although the truncation schemes used there are different. Here we take (2+1)-dimensional SU(2) LGT as an example. The Hamiltonian for this system is:

$$H = \frac{g^2}{2a}[\sum_l E_l^2 - \frac{2}{g^4}\sum_p \mathrm{Tr}(U_p + U_p^\dagger)] \quad . \tag{13}$$

The basic idea of this expansion is to assume the ground state $|\Psi_0\rangle$ and first excited state (the glueball wavefunction) $|\Psi\rangle$ of the Hamiltonian in eq. (13) can be represented by an exponential form:

$$|\Psi_0\rangle = e^{R(U)}|0\rangle$$

$$|\Psi\rangle = F(U)e^{R(U)}|0\rangle \tag{14}$$

where $R(U)$ and $F(U)$ are functions of loop variables and the state $|0\rangle$ is the strong-coupling ground-state, defined by

$$E_l^a|0\rangle = 0 \ . \tag{15}$$

The eigenvalue equation for H can then be written as:

$$\sum_l ([E_l, [E_l, R]] + [E_l, R][E_l, R]) - \frac{2}{g^4}\sum_p \mathrm{Tr}(U_p + U_p^\dagger) = \frac{2a}{g^2}\epsilon_0$$

$$\sum_l ([E_l, [E_l, F]] + 2[E_l, F][E_l, R]) = \frac{2a}{g^2}(\epsilon_1 - \epsilon_0) \tag{16}$$

where ϵ_0 (ϵ_1) is the ground state (the first excited state) energy. $R(U)$ and $F(U)$ can be decomposed according to the order of graphs,

$$R = \sum_i R_i$$

$$F = \sum_i F_i \tag{17}$$

and the lowest order term of R and F is:

$$R_1 = c_1 \ \mathrm{Tr}(U_p + U_p^\dagger) \quad ,$$

$$(18)$$

$$F_1 = f_1 \ \mathrm{Tr}(U_p + U_p^\dagger) \quad .$$

The graphs of order i are generated by the term

$$\sum_{j=1}^{i-1} [E_l, R_j][E_l, R_{i-j}]$$

$$(19)$$

in equation (16).

In order to make the calculation possible, some scheme to truncate the eigenvalue equation must be used. The truncation scheme used by Smith and Watson (1993) is

$$\sum_l ([E_l, [E_l, \sum_{i=1}^{n} R_i]] \ + \sum_{i,j=1}^{n} [E_l, R_i][E_l, R_j]) - \frac{2}{g^4} \sum_p \mathrm{Tr}(U_p + U_p^\dagger) = \frac{2a}{g^2} \epsilon_0$$

$$(20)$$

$$\sum_l ([E_l, [E_l, \sum_{i=1}^{n} F_i]] \ +2 \sum_{i,j=1}^{n} [E_l, F_i][E_l, R_j]) = \frac{2a}{g^2}(\epsilon_1 - \epsilon_0)$$

where the new graphs generated by $[E_l, R_i][E_l, R_j]$ and $[E_l, F_i][E_l, R_j]$ are simply discarded. Guo, Chen and Li (1994) have argued that because of this, the continuum limit of this system may be seriously altered, and they have proposed a better truncation scheme:

$$\sum_l ([E_l, [E_l, \sum_{i=1}^{n} R_i]] \ + \sum_{i+j \leq n} [E_l, R_i][E_l, R_j]) - \frac{2}{g^4} \sum_p \mathrm{Tr}(U_p + U_p^\dagger) = \frac{2a}{g^2} \epsilon_0$$

$$(21)$$

$$\sum_l ([E_l, [E_l, \sum_{i=1}^{n} F_i]] \ +2 \sum_{i+j \leq n} [E_l, F_i][E_l, R_j]) = \frac{2a}{g^2}(\epsilon_1 - \epsilon_0) \quad .$$

The most tedious task for the high-order approximations is to generate a list of independent loop configurations and derive the nonlinear coupled equations. So far, all these calculations in lattice gauge theory have been mainly carried out by hand. We have tried to develop computer algorithms to overcome this problem. Borrowing some ideas from our computer algorithms used to generate a list of clusters for our linked-cluster series expansions and t-expansions (see, for example, Hamer, Zheng and Oitmaa (1996)), a preliminary program was developed.

3 Application to Quantum Hamiltonian Lattice Systems

3.1 The application in 2D Heisenberg antiferromagnet

The $S = \frac{1}{2}$ XXZ Heisenberg antiferromagnet on the square lattice is an interesting example of a quantum Hamiltonian spin model, and has come under intensive study recently because of its possible relevance to high-T_c superconductors. Reviews of this work have been given by Barnes (1991) and Manousakis (1991).

The anisotropic Heisenberg antiferromagnet can be described by the following Hamiltonian:

$$H_0 = \sum_{\langle lm \rangle} [S_l^z S_m^z + x(S_l^x S_m^x + S_l^y S_m^y)] , \qquad (22)$$

where $\langle lm \rangle$ denotes a sum over all nearest-neighbor pairs.

We have recently calculated an extended perturbation series expansion (see Zheng, Oitmaa and Hamer (1991)) and t-expansion (see Zheng, Oitmaa and Hamer (1995)) for the ground state energy E_0/N, the staggered magnetization M^+, and the energy gap m, etc. Witte, Hollenberg and Zheng (1996) have also recently calculated the analytic Lanczos expansion for this system. All these expansions are calculated up to the same lattice size: they all involved a list of 11131 linked-cluster of up to 14 sites for the ground state properties, and a list of 2525 linked and unlinked cluster of up to 11 sites for the excited state. The perturbation series has been analyzed by using integrated differential approximants, and the t-expansion series has been analyzed by using D-Padé approximants, the connected-moment expansion (CMX), the Laplace method and the inversion method. The extended coupled-cluster expansion has been carried out by Zeng, Farnell and Bishop (1996) recently. We have also extended the spin-wave expansion for this system to third order (and fourth order for ground state energy) (see Zheng and Hamer (1993)). The best QMC simulation for this system has been carried out by Runge (1992).

Table 1. A comparison of the ground-state energy E_0/N, the staggered magnetization M^+, and the amplitude of energy gap m at the isotropic point $x = 1$ for the spin-$\frac{1}{2}$ Heisenberg antiferromagnet on a square lattice obtained from spin-wave theory, the linked-cluster series expansion, the t-expansion, the coupled-cluster expansion, and the quantum Monte Carlo simulation.

Method	E_0/N	M^+	$(1 - x^2)^{-1/2}m$
Spin wave theory (Zheng and Hamer (1993))	-0.6693(2)	0.3069	1.24(3)
Series (Zheng, Oitmaa and Hamer (1991))	-0.6693(1)	0.307(1)	1.27(2)
t-expansion (Zheng, Oitmaa and Hamer (1995))	-0.668(1)	0.33(3)	
Analytic Lanczos (Witte, Hollenberg and Zheng (1996))	-0.6691(3)	0.353(5)	
Coupled cluster (Zeng, Farnell and Bishop (1996))	-0.66817	0.3524	
Monte Carlo (Runge (1992))	-0.66934(4)	0.3075(25)	

If we concentrate our attention on the isotropic point $x = 1$, which is physically the most interesting, a comparison of the ground-state energy E_0/N, the staggered magnetization M^+ and the energy gap m at $x = 1$ obtained from different methods are shown in Table 1. First of all, it may be seen that the different methods are generally in good agreement with each other, within errors. We find that the standard perturbation series expansion has given the most accurate results of the linked-cluster methods, being a little less accurate than the best current Monte Carlo estimates for the ground-state energy, but a little more accurate for the magnetization. The spin-wave expansion also gives remarkable accurate results. The coupled cluster expansion, the t-expansion and the analytic Lanczos expansion give a slightly higher estimate for the staggered magnetization.

3.2 The application to 2+1D U(1) LGT

The compact U(1) lattice gauge model in (2+1)D has become a standard proving ground for Hamiltonian lattice numerical techniques. It is one of the simplest models with dynamical gauge field degrees of freedom, and it is similar to QCD in possessing a confining phase for all values of the coupling e^2, as shown by the pioneering studies of Polyakov (1977) and Banks, Myerson and Kogut (1977), and the rigorous analysis of Göpfert and Mack (1982). In the continuum limit it reduces to a massive scalar free field theory, with a mass gap M which is expected (see Göpfert and Mack (1982)) to decrease exponentially as the lattice spacing a goes to zero:

$$M^2 a^2 \sim \frac{c_1}{g^2} \exp(-\frac{c_2}{g^2}) \quad \text{as} \quad a \to 0 \tag{23}$$

where the bare coupling g is related to the charge e by $g^2 = e^2 a$. If one takes the "naive" continuum limit, however, where the coupling e^2 and energy scale are held fixed as a goes to zero, then Gross (1983) has proven that the Villain version of the model, at least, converges to free electromagnetic theory at the level of $F_{\mu,\nu}$ or the Wilson loops, as it was originally designed to do; and a recent studies by Hamer and Zheng (1993) and Hamer, Wang and Price (1994) has shown that this is reflected in the finite-size behaviour of the model.

The U(1) LGT in (2+1)D can be described by the following Hamiltonian:

$$H = \frac{g^2}{2a} \left[\sum_l E_l^2 - x \sum_p (U_p + U_p^\dagger) \right] \tag{24}$$

where the index l labels the links between sites, p labels the elementary plaquettes of the lattice, a is the lattice spacing, $x = 1/g^4$. E_l is the electric field operator at link l, while U_p is the plaquette operator at plaquette p.

In the strong-coupling limit $x \to 0$, the dominant term in the Hamiltonian is the electric field term. The strong-coupling states are therefore eigenstates of

E_l^2 on each link. For the U(1) model, E_l takes integer eigenvalues 0, ± 1, ± 2, ... on each link l; while the plaquette operator

$$U_p = U_1 U_2 U_3^\dagger U_4^\dagger, \tag{25}$$

raises or lowers E_l one unit on links 1, $\cdots$, 4 of the plaquette p, by virtue of the commutation relations:

$$[E_l, U_l] = U_l, \quad [E_l, U_l^\dagger] = -U_l^\dagger. \tag{26}$$

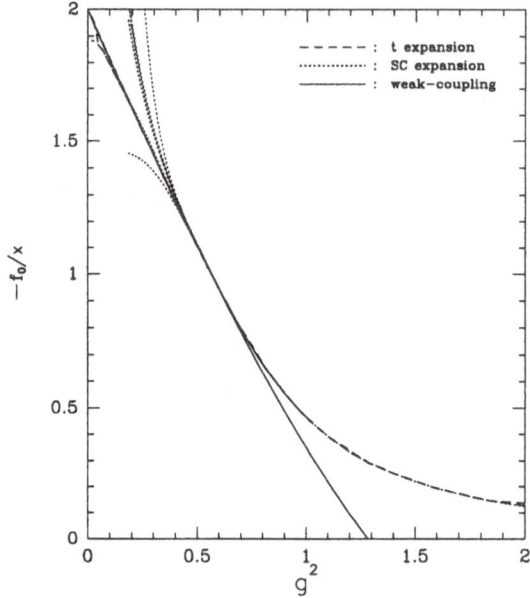

Fig. 1. Graph of $-f_0/x$ against g^2 for the U(1) LGT in (2+1)D. The curves shown are estimates from the t-expansion (obtained from D-Padé approximants) and SC-expansion, and the predictions of a weak-coupling expansion (Hamer and Zheng (1993)).

The Hamiltonian version of this model has been recently studied by a variety of numerical techniques including strong-coupling perturbation series (see Hamer, Oitmaa and Zheng (1992)), weak-coupling perturbation expansion (see Hamer and Zheng (1993)), t-expansions (see Hamer, Zheng and Oitmaa (1996)), coupled cluster expansions (see Fang, Liu and Guo (1995)), the quantum Monte Carlo simulations (see Hamer, Wang and Price (1994)). The results are generally in good agreement with the expected behaviour, such as (23). Here we concentrate on a comparison of the results of three different linked-cluster expansions, where both strong coupling series expansions and t-expansions have been carried out for the ground state energy per site E_0/N, the symmetric and antisymmetric mass gaps (M_S and M_A) and the string tension T_{ax} all up to 9 plaquettes. The coupled-cluster expansion has been carried out for the mass gap and long wavelength wave function by Fang, Liu and Guo (1995) up to 6th order which involved 2456 independent configurations (6 plaquettes).

The SC-series has been analysed by using Shafer approximants and integrated differential approximants, and we use four different methods including D-Padé approximants, the connected-moment expansion (CMX), the Laplace method and the inversion method, to analyze the t-expansion series.

a) **The ground state energy**

Figure 1 shows $-f_0/x$ as a function of g^2 to exhibit the weak coupling behavior. The weak-coupling expansion of the ground state energy f_0 for this model has been calculated by Hamer and Zheng (1993):

$$f_0 = -2x + 1.916183x^{1/2} - 0.2294848 - 0.0268602x^{-1/2} - 0.009315x^{-1} + O(x^{-3/2}) \tag{27}$$

It can be seen from Fig. 1 that the estimates from the t-expansion converge ex-

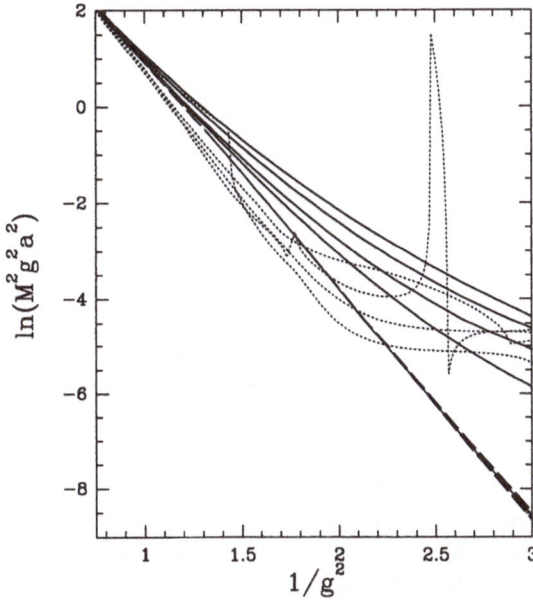

Fig. 2. Graph of $\ln(M_A^2 g^2 a^2)$ against $1/g^2$ for the antisymmetric mass gap. The curves shown are estimates from the t-expansion (dotted lines), SC-expansion (long-dashed lines), and the results of 3 to 6 order (solid lines from top to bottom) coupled-cluster expansions.

tremely well down to coupling of order $g^2 = 0.1$, whereas the estimates from the

SC-expansion only converge well down to coupling $g^2 = 0.5$: beyond that, the approximants begin to spray outwards, being unable to match the true asymptotic behavior obtained from the weak-coupling expansion.

b) **Mass gaps**

Figure 2 shows estimates of the quantity $\ln(M_A^2 g^2 a^2)$ as a function of $1/g^2$ obtained from the three coupled-cluster expansions. The SC-expansion estimates, obtained by the method of integrated differential approximants, show good convergence and follow quite closely the exponential scaling behaviour expected from equation (23). The D-Padé approximants for the t-expansion, on the other hand, do not converge at all well, so that no sensible estimates can be obtained below $g^2 \simeq 0.5$. No other method of extrapolation seems to do any better. The coupled-cluster expansion converges to the of SC expansion as the order increases.

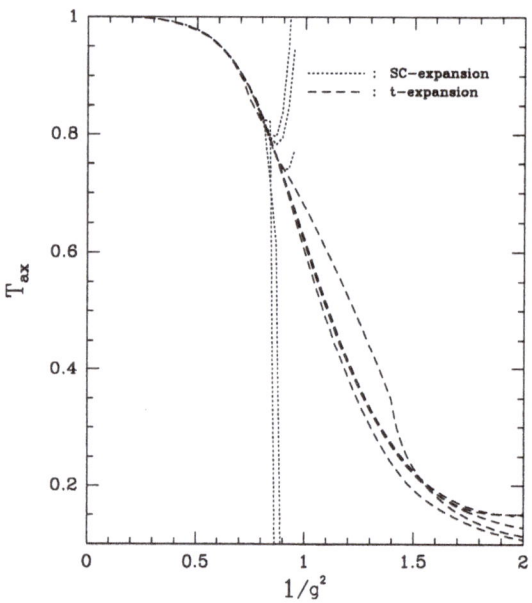

Fig. 3. Graph of the string tension T_{ax} against $1/g^2$. The curves shown are estimates from the t-expansion (obtained from D-Padé approximants) and SC-expansion (obtained from integrated differential approximants).

c) **String tension**

When it comes to the axial string tension, the t-expansion enjoys a crucial advantage. The axial string tension is known to undergo a roughening transition at some finite coupling value, which prevents the analytic continuation of the SC-series beyond that point. This can be seen in Fig. 3, for example, where the SC-series approximants fail to converge beyond $1/g^2 \simeq 0.8$. The t-expansion estimates, on the other hand, do not suffer from the same difficulty, and appear to give sensible results well beyond the roughening transition.

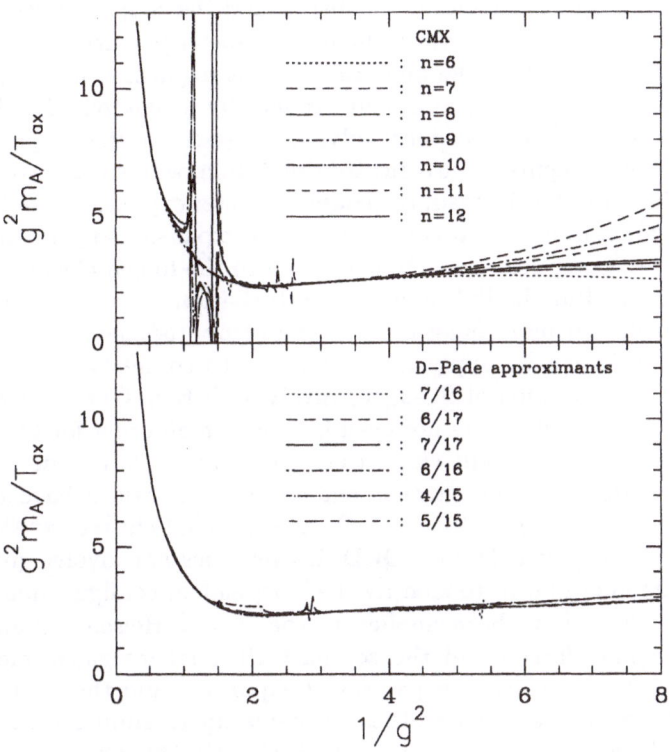

Fig. 4. Graph of the ratio $g^2 m_A/T_{ax}$ against $1/g^2$. The estimates from D-Padé and CMX approximants to the t-expansion series are shown.

The analysis of Göpfert and Mack (1982) predicts that in the weak-coupling limit

$$\frac{e^2 M_A}{\sigma} = \frac{g^2 m_A}{T_{ax}} \to \text{constant} \quad \text{as} \quad g^2 \to 0 \tag{28}$$

where for the Villain version of the model the constant is $\pi^2/2$. Figure 4 displays D-Padé and CMX approximants to the t-expansion for this quantity, as a function of $1/g^2$. There is some noise in the approximants, but they certainly appear to level out beyond $1/g^2 \simeq 2$, at a value in the range $3 - 4$. Note that $g^2 m_A/T_{ax}$ obtained from CMX approximants oscillates with the order n in the large $1/g^2$ region.

4   Summary and Conclusions

Comparisons have been made between several different linked-cluster expansion methods, including the linked-cluster perturbation series expansion, the t-expansion, the analytic Lanczos expansion, and the coupled-cluster expansion.

From a technical point of view, the linked-cluster perturbation series expansion, the t-expansion, and the analytic Lanczos expansion have been well implemented as computer programs, and can usually be computed to higher order (in terms of the size of the cluster involved) compared to the coupled-cluster expansion; but they require some series extrapolation, whereas the coupled-cluster expansion does not, but instead, the coupled-cluster expansion requires the solution of a large set of nonlinear equations. The t-expansion appears to be simpler and require less computer memory, compared to the linked-cluster perturbation series expansion. But the linked-cluster perturbation series expansion has been well developed, and more quantities can be computed such as the full excitation spectrum. When we apply these methods to complicated system such as the QCD, the calculation of SC-series involves Clebsch-Gordan coefficients, or at least $6j$ symbols, which are not even known at high order for the SU(3) case. The t-expansion is technically easier to calculate for such a model, since it only involves the SU(3) group integration, and a number of results have already been obtained for QCD by Schreiber (1994). The coupled-cluster expansion is also technically easier to calculate for QCD, it even does not involve group integration, but instead one needs to identify the independent configurations generated.

These methods have been applied to the $S = \frac{1}{2}$ Heisenberg antiferromagnet on the square lattice and the compact U(1) lattice gauge model in 2+1 dimensions (where the series expansion, t-expansion and the analytic Lanczos expansion have been applied to these systems up to similar lattice size). We found that the series expansion gives the best estimates for the $S = \frac{1}{2}$ Heisenberg antiferromagnet on the square lattice, whereas other methods give higher estimates for the staggered magnetization. For the U(1) lattice gauge model in 2+1 dimensions, it has been found that the t-expansion and SC-series estimates are both quite accurate for the ground-state energy, with the t-expansion probably having a slight advantage; whereas for the mass gaps, the SC-expansions are definitely superior for this model. For the axial string tension, no results have been presented for the axial string tension from the coupled-cluster expansion as far as we know, and the SC-expansion cannot be continued beyond the roughening transition, and provides no estimate of the weak-coupling behaviour, whereas the t-expansion estimates can be continued quite happily.

Our conclusion then is that all these methods have their advantages in different contexts, and all will continue to have important applications in quantum Hamiltonian lattice system. The perturbation series expansion is the preferable technique for quantum spin systems, whereas for non-Abelian LGT such as SU(3) Yang-Mills theory, the t-expansion and the coupled-cluster expansion is the preferable technique, although a more reliable method of estimation for the mass gaps would be desirable.

This work forms part of a research project supported by a grant from the Australian Research Council.

References

Banks, T., Myerson, R. and Kogut, J. (1977): Phase transitions in Abelian lattice gauge theories. Nucl. Phys. B**129**, 493-510.

Barnes T. (1991): The 2D Heisenberg antiferromagnetic in high-T_c superconductivity, a review of numerical techniques and results. Int. J. Mod. Phys. C**2**, 659-709.

Bishop, R.F. (1991): An overview of coupled cluster theory and its applications in physics. Theor. Chim. Acta **80**, 95-148.

Coester, F., and Kümmel, H. (1960): Short-range correlations in nuclear wave functions. Nucl. Phys. **17**, 477-485.

Domb, C. and Green, M.S. (1974): Phase Transitions and Critical Phenomena. Vol. 3.

Fang, X.Y., Liu, J.M. and Guo, S.H. (1996): Vacuum wave function and mass gaps of U(1) lattice gauge theory in 2+1 dimensions. Phys. Rev. D **53** 1523-1527;

Gelfand, M.P. (1996): Series expansions for excited states of quantum lattice models, Solid State Communications, **98**, 11-14.

Gelfand, M.P., Singh, R.R.P. and Huse, D.A. (1990): Perturbation expansions for quantum many-body systems. J. of Stat. Phys. **59**, 1093-1142.

Gross, L.G. (1983): Convergence of U(1)$_3$ lattice gauge theory to its continuum limit. Commun. Math. Phys. **92**, 137-162.

Guo, S.H., Chen, Q.Z. and Li, L. (1994): Analytic calculation of the vacuum wave function for (2+1)-dimensional SU(2) lattice gauge theory. Phys. Rev. D **49**, 507-510.

Göpfert, M. and Mack, G. (1982): Proof of confinement of static Quarks in 3-dimensional U(1) lattice gauge theory for all values of the coupling constant. Commun. Math. Phys. **82**, 545-606.

Hamer, C.J., Oitmaa, J. and Zheng, W.H. (1992): Series analysis of U(1) and SU(2) lattice gauge theory in 2+1 dimensions. Phys. Rev. D **45**, 4652-4658.

Hamer, C.J., Wang, K.C. and Price, P.F. (1994): Finite-size scaling for the U(1) lattice gauge model in 2+1 dimensions. Phys. Rev. D **50**, 4693-4702.

Hamer, C.J. and Zheng, W.H. (1993): Weak-coupling expansions and effective lagrangian for compact U(1) lattice gauge theory in D+1 dimensions. Phys. Rev. D **48**, 4435-4449.

Hamer, C.J., Zheng, W.H. and Oitmaa, J. (1994): Spin-wave stiffness of the Heisenberg antiferromagnet at zero temperature. Phys. Rev. B **50**, 6877-6888.

Hamer, C.J., Zheng, W.H. and Oitmaa, J. (1996): Comparison between linked-cluster expansion methods for the U(1) lattice gauge model in 2+1 dimensions. Phys. Rev. D **53**, 1429-1438.

He, H.X., Hamer, C.J. and Oitmaa J. (1990): High-temperature series expansions for the (2+1)D Ising model. J. Phys. A **23**, 1775-1787.

Hollenberg, L.C.L. and Witte, N.S. (1994): General nonperturbative estimate of the energy density of lattice Hamiltonians. Phys. Rev. D **50**, 3382-3386.

Horn, D. and Weinstein, M., (1984): The t expansion: a nonperturbative analytic tool for Hamiltonian systems. Phys. Rev. D **30**, 1256-1270.

Irving, A.C. and Hamer, C.J., (1984): Methods in Hamiltonian lattice field theory (II) Linked-cluster expansions. Nucl. Phys. B**230**, 361-384.

Nickel, B.G. (1980): unpublished.

Manousakis E. (1991): The spin-$\frac{1}{2}$ Heisenberg antiferromagnet on a square lattice and its application to the Cuprous Oxides. Rev. Mod. Phys. **63**, 1-62.

Marland, L.G., (1981): Series expansions for the zero-temperature transverse Ising model. J. Phys. A **14**, 2047-2057.

Morningstar, C.J. (1992): Bistate t-expansion study of U(1) lattice gauge theory in 2+1 dimensions. Phys. Rev. D**46**, 824-835.

Polyakov, A.M. (1977): Quark confinement and topology of gauge theories. Nucl. Phys. B**120**, 429-458.

Runge, K.J. (1992): Finite-size study of the ground state energy, susceptibility, and spin-wave velocity for the Heisenberg antiferromagnet. Phys. Rev. B **45**, 12292-12296.

Schreiber, D. (1994): t-expansion of heavy-light mesons, Phys. Rev. D **49**, 2567-2573.

Smith, C.H.L and Watson, N.J (1993): The shifted coupled cluster method. A new approach to Hamiltonian lattice gauge theories. Phys. Letts. B**302**, 463-471.

Witte, N.S., Hollenberg, L.C.L. and Zheng, W.H. (1996): 2D XXZ model ground state Properties using an analytic Lanczos expansion. submitted to Phys. Rev. B.

Zeng, C., Farnell, D.J.J. and Bishop, R.F. (1996): An efficient implementation of high-order coupled-cluster techniques applied to quantum magnets. cond-mat/9611012.

Zheng, W.H. and Hamer, C.J. (1993): Spin-wave theory and finite-size scaling for the Heisenberg antiferromagnet. Phys. Rev. B **47**, 7961-7970.

Zheng, W.H., Oitmaa, J. and Hamer, C.J. (1991): The square-lattice Heisenberg anti-ferromagnet at $T = 0$. Phys. Rev. B **43**, 8321-8330.

Zheng, W.H., Oitmaa, J., and Hamer, C.J. (1995): Comparison between linked-cluster expansion methods for the Heisenberg antiferromagnet on the square lattice. Phys. Rev. B **52**, 10278-10285.

Critical Properties of 1-D Spin 1/2 Antiferromagnetic Heisenberg Model

J.F. Audet[1], A. Fledderjohann[2], C. Gerhardt[2], M. Karbach[2], H. Kröger[1]*, K.H. Mütter[2] and M. Schmidt[2]

[1] Département de Physique, Université Laval, Québec, Québec, G1K 7P4, Canada
   Email: hkroger@phy.ulaval.ca
[2] Physics Department, University of Wuppertal, D-42097 Wuppertal, Germany Email: muetter@wptsb.physik.uni-wuppertal.de

Abstract. We discuss numerical results for the $1 - D$ spin 1/2 antiferromagnetic Heisenberg model with next-to-nearest neighbour coupling in the presence of a uniform magnetic field. The model develops zero frequency excitations at field-dependent soft-mode momenta. We compute critical quantities from finite size dependence of static structure factors.

1 Introduction

The motivation for studying quantum spin systems in one space dimension (spin chain) can be laid out as follows: There are models which are exactly soluble, e.g., the Heisenberg model. "Exact" numerical results can be obtained by diagonalization of the Hamiltonian for reasonably large chains (up to approximately 50 spins). These quite precise data allow one to carry out a finite size analysis and to extrapolate it to the thermodynamic limit. On the other hand, $1 - D$ spin chains have been analyzed by different techniques, like quantum Monte Carlo methods, the coupled cluster method ($exp(S)$-method), renormalization group techniques, etc. They thus constitute a benchmark problem for comparisons of various methods. From the physics point of view, some substances occuring in nature manifest predominantly the structure of a $1 - D$ chain. Examples include $CuCl_2 2NC_5 H_5$ and $Cu(NH_3)_4 SO_4 H_2 O$ (Mazzi 1955), (Dunitz 1957), (Bernasconi 1981). New neutron scattering experiments on the quasi 1-D antiferromagnet $KCuF_3$ are discussed by Tennant et al. (Tennant 1995). In recent investigations Dender et al. (Dender 1996) have considered copper benzoate, also being a linear chain antiferromagnet. According to Inui(Inui 1988), the Heisenberg model has a relation to the Hubbard model: the pair hopping term of the Hubbard model can be mapped onto the next-to-nearest neighbour interaction of the antiferromagnetic Heisenberg (AFH) model. $1 - D$ spin chains are theoretically interesting because their critical properties are predicted by conformal symmetry, under the hypothesis that this symmetry holds. Finally, $1 - D$ spin chains in the presence of magnetic fields and next-to nearest neighbour coupling display a rich phase

---

* talk presented by H. Kröger

structure. The purposes of this paper are to consider $1-D$ spin chains, to present numerical results from exact diagonalization, and to discuss their finite-size analysis and critical behaviour. We mainly consider static spin structure factors. In particular, we analyze chains with next-to-nearest neighbour coupling, which allows for frustration.

2 Heisenberg model and its extensions

Let us start by recalling the definition of various $1-D$ spin models. A classical model, describing next-neighbour interaction is the Ising model

$$H = J \sum_{i=1}^{N} S_i^z S_{i+1}^z, \tag{1}$$

where S_i^z is the z-component of spin i and takes the values $-1, +1$. A quantum model is the Heisenberg model,

$$H = J \sum_{i=1}^{N} \mathbf{S}_i \cdot \mathbf{S}_{i+1}, \tag{2}$$

where $\mathbf{S} = \sigma/2$ and σ_a represents the Pauli-matrices for the spin $1/2$ case. There is also a classical Heisenberg model with classical 3-component spins. The quantum Heisenberg model is considered the standard model of magnetism. When $J > 0$ the model describes antiferromagnetism, while $J < 0$ describes the ferromagnetic system. The following generalizations are of physical interest: The XXZ model describes the Heisenberg model with an anisotropy in z-direction,

$$H = J \sum_{i=1}^{N} S_i^x S_{i+1}^x + S_i^y S_{i+1}^y + cos(\gamma) S_i^z S_{i+1}^z. \tag{3}$$

The Heisenberg model exposed to an external magnetic field is given by

$$H = J_1 \sum_{i=1}^{N} \mathbf{S}_i \cdot \mathbf{S}_{i+1} + J_B \sum_{i=1}^{N} \mathbf{B}_i \cdot \mathbf{S}_i. \tag{4}$$

If one takes into account next-to-nearest neighbour coupling of spins, the model becomes

$$H = J_1 \sum_{i=1}^{N} \mathbf{S}_i \cdot \mathbf{S}_{i+1} + J_2 \sum_{i=1}^{N} \mathbf{S}_i \cdot \mathbf{S}_{i+2}, \quad \alpha = J_2/J_1. \tag{5}$$

If the signs of J_1 and J_2 are such that one term favours parallel spins, while the other favours antiparallel spins, the system is said to become frustrated. A model of particular interest with coupling beyond next neighbours is the

Haldane-Shastry model (Haldane 1988), (Shastry 1988) (with periodic boundary conditions),

$$H = J \sum_{n=1}^{N-1} \sum_{m=1}^{N} \frac{1}{\sin^2(n\pi/N)} \mathbf{S}_m \cdot \mathbf{S}_{m+n}. \tag{6}$$

Below we will consider combinations of the above cases, e.g., the presence of a magnetic field and next-to-nearest neighbour coupling. In the subsequent sections we will consider spin models with periodic boundary conditions.

2.1 Symmetries and integrability of Heisenberg model

The isotropic Heisenberg model displays a number of symmetries and corresponding conserved quantities: First, there is translational symmetry. The translation operator $T = \exp[iP]$ defines the conserved quantum number of momentum P. Next, there is reflexion symmetry. The total spin $\mathbf{S} = \sum_{i=1} \mathbf{S}_i$ is conserved. There are more conserved quantities like, e.g., $F_3 = 2i \sum_{n=1}^{N} \epsilon_{ijk} S_{n-1}^i S_n^j$ S_{n+1}^k (Lüscher 1976), (Fabricius 1990). The existence of conserved quantities can be traced back to the existence of a set of commuting transfer matrices $T(\lambda)$, satisfying $[T(\lambda), T(\lambda')] = 0$ and $[T(\lambda), H] = 0$. The existence of such transfer matrices for the $1 - D$ Heisenberg model can be proved either from inverse scattering theory or via the Yang-Baxter equations (Baxter 1980). The set of transfer matrices implies the existence of infinitely numerous conserved quantities like, e.g., the Hamiltonian $H \propto \frac{\partial}{\partial \lambda} \ln T(\lambda) \big|_{\lambda=1/2} + constant$, the total spin squared $\mathbf{S}^2 \propto \frac{\partial^2}{\partial \lambda^2} T(\lambda) \big|_{\lambda=\infty}$, or $F_3 \propto \frac{\partial^2}{\partial \lambda^2} \ln T(\lambda) \big|_{\lambda=i/2} + constant$ (Fabricius 1990). Thus, the Heisenberg model is an integrable system (Baxter 1980). Analytical solutions have been obtained via the Bethe ansatz (Bethe 1931), with which Hulthén (Hulthén 1938) has calculated the ground state energy per site in the thermodynamic limit $\epsilon_0 = 1/4 - \ln 2$ (corresponding to $J = 1/2$). Analytical information on the $1 - D$ model at the critical point can be obtained by making the assumption of conformal invariance. Cardy (Cardy 1986), (Cardy 1987) has shown that the spectrum of the transfer matrix is determined by conformal invariance. This predicts, for example, the leading term of the finite size behaviour of the ground state energy, or critical exponents of the spin-spin correlation function.

Recently, significant analytic progress has been made in the Heisenberg and related spin models. Römer and Sutherland (Römer 1993) have computed the finite size dependence of the correlation function of the AFH model. The dynamic correlation function of the Calogero-Sutherland model has been computed by Ha (Ha 1994) while the single-particle Green's function of the same model has been determined by Zirnbauer and Haldane (Zirnbauer 1995). Essler et al. (Essler 1996) have further shown that correlation functions of the XXZ antiferromagnet near the critical magnetic field can be expressed in terms of solutions of Painlevé differential equations.

2.2 Quantum numbers of lowest state for finite chains

The quantum numbers of the lowest lying state for a finite spin chain have been analyzed by Fabricius et al. (Fabricius 1991). For a given total spin S and number of spins N one finds the quantum number of momentum P of the lowest lying state to be as follows: For N even it holds that

$$P(\text{mod } 2\pi) = -\pi \times \begin{cases} S : N = 4, 8, 12, \cdots \\ S + 1 : N = 2, 6, 10, \cdots \end{cases} \tag{7}$$

For N odd, there is degeneracy of momentum

$$P^+(\text{mod } 2\pi) = -\pi \times \begin{cases} S(1 + 1/N) : N = 3, 7, 11, \cdots \\ S(1 + 1/N) + 1 : N = 5, 9, 13, \cdots \end{cases}$$

$$P^-(\text{mod } 2\pi) = -\pi \times \begin{cases} S(1 - 1/N) + 1 : N = 3, 7, 11, \cdots \\ S(1 - 1/N) : N = 5, 9, 13, \cdots \end{cases} \tag{8}$$

The absolute ground state corresponds to

$$P(S = 0) = \begin{cases} 0 : N = 4, 8, 12, \cdots \\ \pi : N = 2, 6, 10, \cdots \end{cases}$$

$$P^\pm(S = 1/2) = \mp\pi/2 \times \begin{cases} (1 + 1/N) : N = 3, 7, 11, \cdots \\ (-1 + 1/N) : N = 5, 9, 13, \cdots \end{cases} \tag{9}$$

These are the quantum numbers corresponding to the isotropic Heisenberg model. In the case of next-to-nearest neighbour coupling, the situation becomes more complicated. Then the quantum numbers of the lowest lying state will depend upon the specific model parameters, like α.

2.3 Magnetization curve and susceptibility

Magnetization defined by $\mathbf{M} = < \mathbf{S} > /N$, is a function of the external field $\mathbf{B}$. The functional dependence M versus B, the so-called magnetization curve, is most easily obtained by computing the ground state energy ϵ per spin in a given spin sector and using the equivalence

$$\frac{\partial \epsilon(M)}{\partial M} = B. \tag{10}$$

This follows from

$$H = H|_{B=0} + \mathbf{B} \cdot \mathbf{S},$$
$$E = E|_{B=0} + \mathbf{B} \cdot < \mathbf{S} >,$$
$$\epsilon = E/N = \epsilon|_{B=0} + \mathbf{B} \cdot \mathbf{M}. \tag{11}$$

The magnetization curve of the AFH with anisotropy at temperature zero is shown in Fig.[1]. One observes that when the magnetic field B reaches some

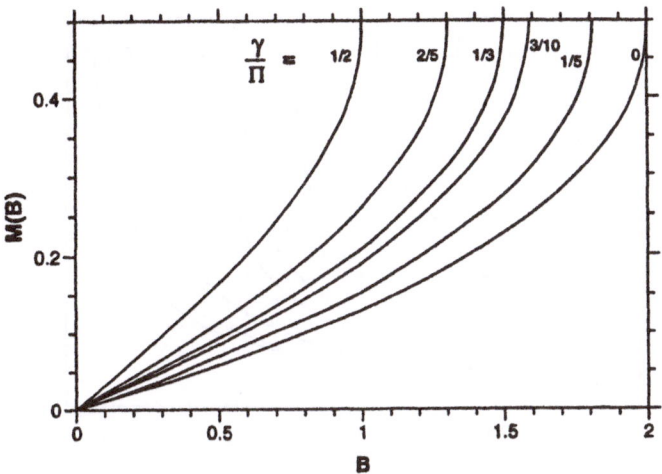

Fig. 1. Magnetization curve computed from Bethe ansatz equations for chain of length N=2048. γ is the anisotropy parameter. Taken from (Karbach 1994a).

critical value B_c, the magnetization reaches its maximal value $M = 1/2$ (saturation, all spins are aligned). This can be understood as follows: For very long chains, i.e., near the thermodynamic limit, and close to saturation (nearly all spins aligned), the change in energy when flipping a single spin becomes independent of magnetization M. Thus $B = \partial\epsilon/\partial M = $ constant and B does not increase as M increases. The magnetization curve for the isotropic AFH model is analytically known to leading order in two limiting cases: (a) weak magnetic field and (b) near saturation. The case of a weak magnetic field has been computed by Griffith (Griffith 1964) as well as by Yang and Yang (Yang 1966) using the Bethe ansatz and solving integral equations numerically. One obtains the expansion

$$M(B) = \frac{B}{\pi}\left[1 - a_1\frac{1}{\ln B} - a_2\frac{\ln|\ln B|}{\ln^2 B} - a_3\frac{1}{\ln^2 B} + \cdots\right]. \qquad (12)$$

The lowest order is due to Griffith (Griffith 1964). The coefficient $a_1 = 1/2$ has been computed by Babujian (Babujian 1983), while $a_2 = 1/4$ has been obtained by Lee and Schlottmann (Lee 1987), but a_3 is unknown. The magnetization curve near saturation is known from the work of Yang and Yang (Yang 1966),

$$M(B) = \frac{1}{2} - \frac{\sqrt{2}}{\pi}\sqrt{B_c - B} + O(B_c - B), \qquad (13)$$

and $B_c = 2$ for the isotropic case. From the magnetization curve, one obtains the magnetic susceptibility $\chi = \frac{\partial}{\partial B}M(B)$. The susceptibility corresponding to the magnetization curve of Fig.[1] is shown in Fig.[2]. Information on antiferromagnetic ordering is encoded in the ground state spin-spin correlation function $< 0|\mathbf{S}(x)\cdot\mathbf{S}(0)|0 >$. For spin chains being isotropic with respect to coupling of x-

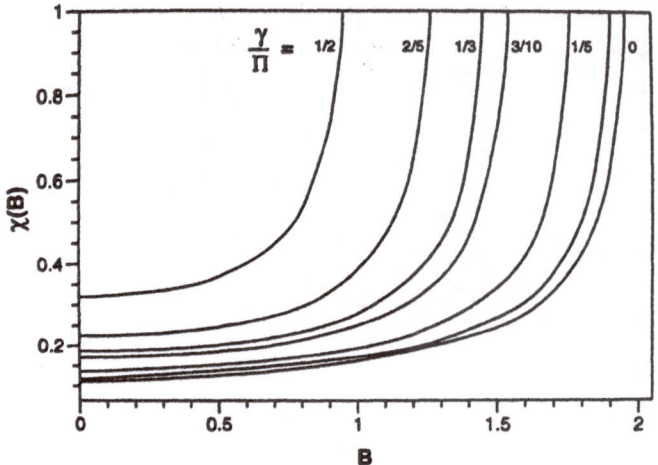

Fig. 2. Susceptibility computed from Bethe ansatz equations for chain of length N=2048. γ is the anisotropy parameter. Taken from (Karbach 1994a).

and y-components of spin, one distinguishes the transverse and the longitudinal correlation function,

$$\omega_j(l,N) = 4 < 0|S_{l+1}^j S_1^j|0 >, \quad j = 1,2 : \text{transverse}, \; j = 3 : \text{longitudinal.} \quad (14)$$

Luther and Peschel (Luther 1975) have found the following behaviour of the correlation functions in the thermodynamic limit

$$\omega_j(l,N) \sim_{N\to\infty} \frac{(-1)^l}{l^{\eta_j}}. \quad (15)$$

For the isotropic Heisenberg model, the critical exponents η_j obey $\eta_j = 1$ for $j = 1,2,3$. This follows from conformal symmetry (Luther 1975), (Bogoliubov 1986). Also based on conformal symmetry, Affleck et al. (Affleck 1989) have later predicted for the isotropic Heisenberg model a logarithmic correction to the long range behaviour,

$$< S(x)S(0) > \sim A\frac{(-1)^x}{x}(\ln x)^\sigma, \quad \sigma = 1/2. \quad (16)$$

3 Finite size scaling

Critical behaviour of a physical theory can exist only in the thermodynamic limit $(N \to \infty)$. Because very few models are soluble at this limit, it is of great practical interest if physical information of the critical system can be obtained from a finite number of spins (finite volume). This need has been met by the theory of

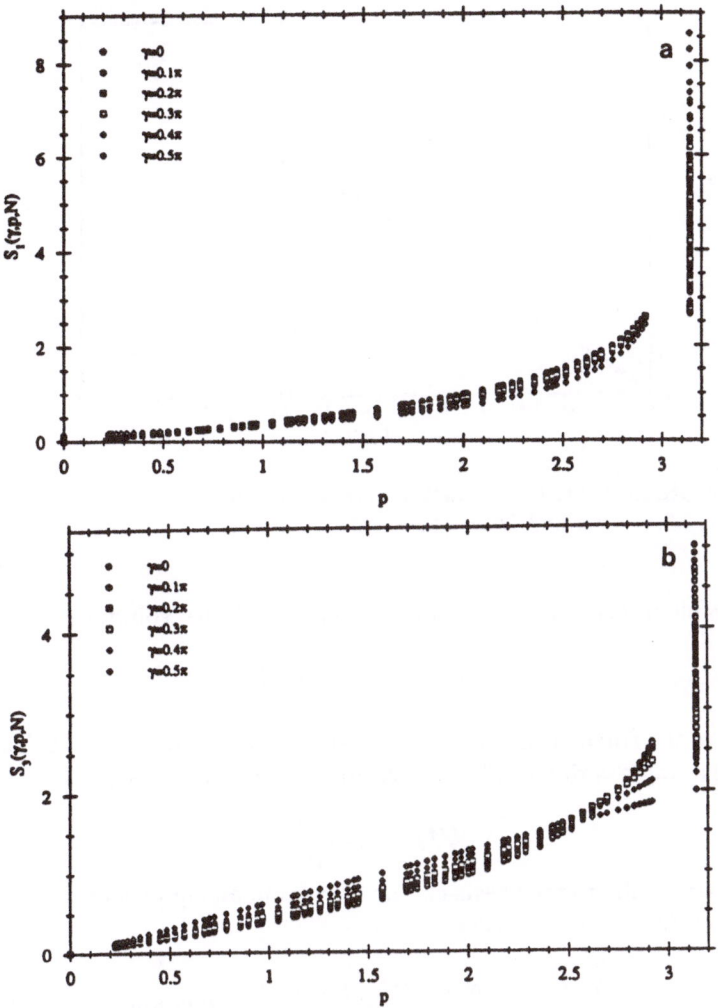

Fig. 3. Transverse (a) and longitudinal (b) structure factor for $N = 4, \cdots, 28$ and different anisotropy parameters γ. Taken from (Karbach 1994a).

finite size scaling, introduced by Fisher (Fisher 1971). It can be summarized as follows: Suppose a model becomes critical at a temperature T_c. For the reduced temperature $t = (T - T_c)/T_c$ this corresponds to $t_c = 0$. Suppose further that there is an observable, which at the critical point behaves according to a power law

$$\Gamma(t) \longrightarrow_{t \to 0} \frac{1}{t^\phi}, \tag{17}$$

where ϕ is a critical exponent. In a finite volume N the observable is $\Gamma(t, N)$. The theory of finite-size scaling is based on the following assumption relating

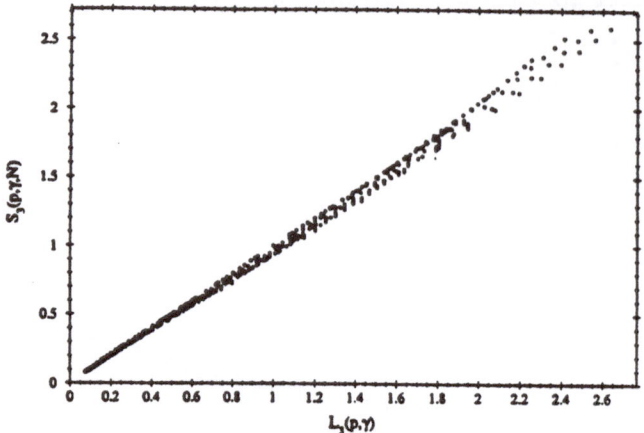

Fig. 4. Longitudinal structure factor versus scaling variable L_3, Eq.(28), for $\gamma/\pi = 0, 0.1, 0.2, \cdots, 0.5$ and $N = 4, 6, \cdots, 28$.

the observable in the thermodynamic limit to the finite volume:

$$\Gamma(t, N) \longrightarrow_{N \to \infty} \Gamma(t) f\left(\frac{N}{\xi(t)}\right), \tag{18}$$

where $\xi(t)$ is the correlation length in the thermodynamic limit and $f(z)$ is some function of a single variable. The correlation length manifests a critical behaviour

$$\xi(t) \longrightarrow_{t \to 0} \frac{1}{t^\nu}. \tag{19}$$

The finite-size scaling hypothesis can be written in an equivalent form, originally proposed by Fisher (Fisher 1971),

$$\Gamma(t, N) \longrightarrow_{N \to \infty} N^{\phi/\nu} g(Nt^\nu)\Big|_{t=\text{constant}}. \tag{20}$$

According to a theorem by Mermin and Wagner (Mermin 1966) the Heisenberg model in $D = 1$ and $D = 2$ dimensions does not have long range order and hence does not have a second order phase transition for any finite temperature $T \neq 0$. The critical exponent ν of the correlation length has been computed in the thermodynamic limit for the XXZ spin model at the critical point $T_c = 0$ by Suzuki et al. (Suzuki 1992), using the Bethe ansatz. The result is $\nu = 1$.

4 Finite size analysis of spin-spin structure factors

Although the Bethe ansatz allows in principle for an exact calculation of the spin-spin correlation function; in practice, it is very difficult. As an alternative, many workers have numerically computed correlation functions and structure factors. The following standard methods have been successfully applied to

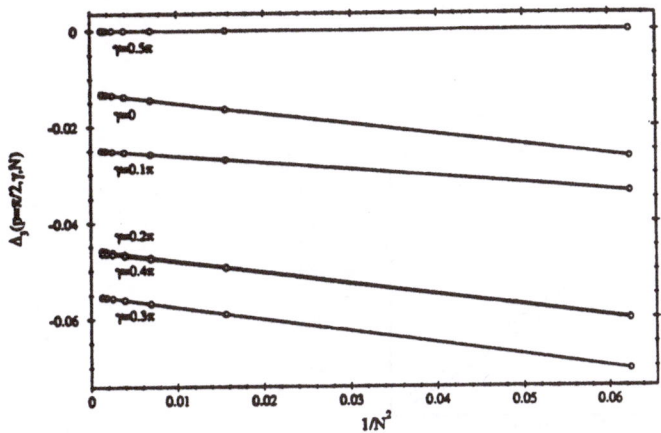

Fig. 5. Finite-size dependence of $\Delta_j(\gamma, p, N)$, Eq.(31), at momentum $p = \pi/2$.

spin systems: exact diagonalization (Bonner 1964), coupled cluster - or $\exp(S)$ - method (Bishop 1995), the quantum Monte Carlo method (Ceperley 1995), and renormalization group techniques (White 1993), (Hallberg 1995). In the following sections we will focus on results obtained by exact diagonalization. Due to limitations in computing time and storage, exact diagonalization is possible only for small systems (for spin 1/2 chain of N spins in 1-D, the dimension of Hilbert space is 2^N). However, in the presence of a strong magnetic field B close to saturation, nearly all spins are aligned. Setting up a Hilbert space basis by counting configurations of non-aligned (reversed) spins, the dimension will be much smaller. E.g., for $N = 50$ spins in the sector of magnetization $M = 2/5$, the dimension of Hilbert space is $d = 42376$. Thus spin chains of up to 56 spins have been diagonalized, using the Lançzos method and binary coding of spin configurations. The advantage of exact diagonalization is its high numerical precision which allows one to carry out a finite size analysis and to extrapolate to the thermodynamic limit.

4.1 Structure factors in momentum space

Physical information on magnetic ordering is encoded in the spin-spin correlators. Numerically, however, it is easier to consider the Fourier transform of the correlation function, i.e., the spin structure factor in momentum space (Karbach 1993), instead of analyzing the long range behaviour of the correlation function in configuration space. The static spin structure factor is defined by

$$S_j(p, N) = \sum_{l=0}^{N-1} \omega_j(l, N) \exp[ipl], \quad p = 2\pi k/N, \quad k = 0, \cdots, N-1. \quad (21)$$

50

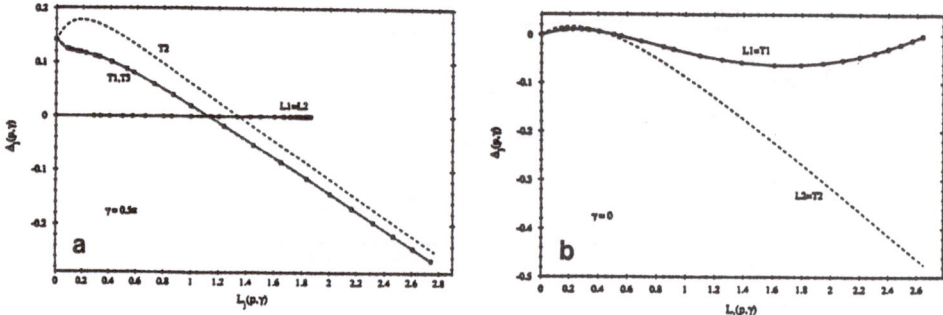

Fig. 6. Thermodynamic limit of transverse (T) and longitudinal (L) structure factor: $T1$, $L1$ finite-size analysis, $T2$, $L2$ conjecture of Ref. (Müller 1981), $T3$ exact result of XX-model. (a) $\gamma = \pi/2$, (b) $\gamma = 0$.

One has the following correspondence

$$\text{x-space: } <S(x)S(0)> \sim \frac{(-1)^x}{x},$$

$$\text{p-space: } <S(p)> \sim \sum_x e^{ipx}\frac{(-1)^x}{x} = \sum_x \frac{e^{i(p+\pi)x}}{x} \quad \text{divergent at } p = \pm\pi \quad (22)$$

Thus, the long distance behaviour of the correlator manifests itself as divergence in p-space. Such behaviour can be analyzed numerically for finite chains.

4.2   Results for XXZ-model

For the XXZ-model with anisotropy angle γ, Eq.(3), the correlation function behaves as (Luther 1975)

$$\omega_j(\gamma, l) \sim c(\gamma)\frac{(-1)^l}{l^{\eta_j(\gamma)}}, \quad \eta_1(\gamma) = \eta_3^{-1}(\gamma) = 1 - \gamma/\pi, \quad 0 \leq \gamma \leq \pi/2. \quad (23)$$

Note that $\eta_1(\gamma) \cdot \eta_3(\gamma) = 1$. The corresponding structure factor behaves as

$$S_j(\gamma, p) \sim a_j + b_j \left[1 - p/\pi\right]^{\eta_j - 1}. \quad (24)$$

Because $\eta_1(\gamma) < 1$ and $\eta_3(\gamma) > 1$, at momentum $p = \pi$ the transverse structure factor becomes divergent, while the longitudinal structure factor remains finite.

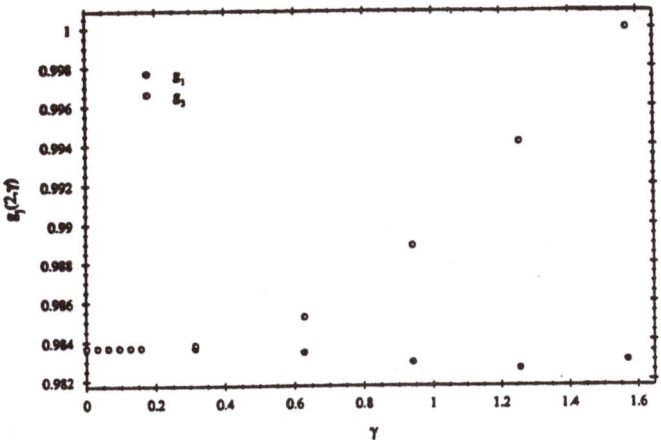

Fig. 7. Scaling function g_j in the finite-size scaling ansatz of Eq.(32) for $z = 2$.

This behaviour changes when $\gamma \to 0$ (isotropic limit). For finite systems the corresponding correlation function and structure factor are given by

$$\omega_j(\gamma, l) \sim \bar{c}_j(\gamma) \frac{(-1)^l}{l^{\eta_j(\gamma)}} \exp\left[l/\xi_j(\gamma, N)\right],$$
$$S_j(\gamma, p) \sim \tilde{a}_j + \tilde{b}_j \left[(1 - p/\pi)^2 + 1/\xi_j^2(\gamma, N)\right]^{\eta_j - 1}, \tag{25}$$

where $\xi_j(\gamma, N)$ is the correlation length in the finite system. The correlation length has the following critical behaviour when $\cosh(\gamma') > 1$ and $\gamma' \to 0$ (Johnson 1973)

$$\xi \longrightarrow \frac{1}{8} \exp\left[\frac{\pi^2}{2\gamma'}\right] \quad \text{divergent at } \gamma' = 0. \tag{26}$$

The structure factor diverges at $p = \pi$ when $\gamma' \to 0$ (Singh 1989)

$$S_j(\gamma', p = \pi, N = \infty) \xrightarrow{\gamma' \to 0} \frac{1}{\gamma'^{\lambda_j}}, \quad \lambda_1 = 3/2, \lambda_3 = 2. \tag{27}$$

The longitudinal and transverse structure factors for finite systems are shown in Fig.[3]

Now let us look at the scaling properties of the structure factors. One can ask the following questions: (a) Does a scaling variable L_j exist such that $S_j(\gamma, p, N) = S_j(\gamma, L_j(\gamma, p, N))$? (b) How does $S_j(\gamma = 0, p, N = \infty)$ behave when $p \to \pi$? (c) What is the leading finite-size behaviour of $S_j(\gamma, p, N)$ when $p < \pi$? (d) How does $S_j(\gamma, p = \pi, N)$ behave when N is large and $\gamma \to 0$? Karbach and Mütter (Karbach 1993) have suggested using $L_3(\gamma, p)$ as a scaling variable, given by

$$L_3(\gamma, p) = \frac{\eta_3(\gamma)}{\eta_3(\gamma) - 1} \left[1 - (1 - p/\pi)^{\eta_3(\gamma) - 1}\right], \quad p < \pi. \tag{28}$$

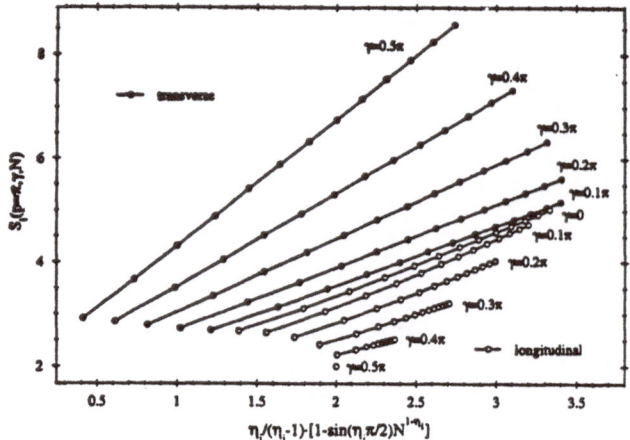

Fig. 8. Scaling behaviour of transverse structure factor at $p = \pi$, compared with finite-size scaling ansatz of Eq.(33).

Haldane and Shastry (Haldane 1988), (Shastry 1988) have shown that

$$\lim_{\gamma \to 0} L_3(\gamma, p) = -\ln(1 - p/\pi), \quad p < \pi \tag{29}$$

is the exact structure factor of the Haldane-Shastry model, Eq.(6), with periodic boundary conditions. Numerical studies (Karbach 1993), (Karbach 1994b) show the following scaling behaviour. Fig.[4] gives a plot of the longitudinal structure factor versus the scaling variable L_3. The data points lie close to a line of slope one when $p < \pi$. Similar but less impressive scaling holds for the tranverse component. Scaling deviations can be further analyzed by looking at

$$\Delta_j(\gamma, p, N) = S_j(\gamma, p, N) - L_j(\gamma, p). \tag{30}$$

It is remarkable that when $\gamma = \pi/2$ (XX-model), no finite-size dependence exists, such that $S_3(\gamma = \pi/2, p, N) = 2p/\pi$. For $p < \pi$ one makes the following finite size ansatz

$$\Delta_j(\gamma, p, N) = \Delta_j(\gamma, p, N = \infty) + \frac{c_j(\gamma, p)}{N^2}, \quad p < \pi. \tag{31}$$

The finite size behaviour is shown in Fig.[5] and confirms the leading $1/N^2$ behaviour when approaching the thermodynamic limit. The case $\gamma = \pi/2$ corresponds to the exactly soluble XX-model. Fig.[6] shows the thermodynamic limit of the structure factors. For $\gamma = \pi/2$ comparison with the exact result shows agreement. The above analysis is not valid when approaching $p = \pi$. The following finite-size scaling ansatz can be made in the combined limit $N \to \infty$, $p \to \pi$,

$$S_j(\gamma, p, N) \xrightarrow{p \to \pi, N \to \infty} S_j(\gamma, p, N = \infty) \, g_j(\gamma, z)|_{z=(1-p/\pi)N=\text{constant}} \tag{32}$$

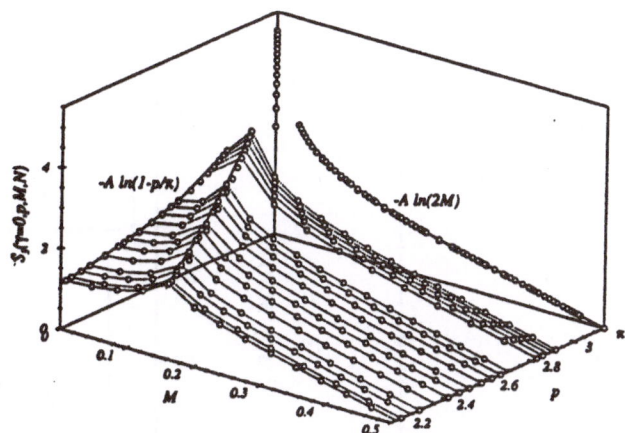

Fig. 9. Longitudinal structure factor versus momentum p and magnetization M for $N = 20, 22, \cdots, 28$. The ridge occurs at soft-mode $p_3(M)$, Eq.(34).

The scaling function g_j is shown in Fig.[7]. The transverse structure factor is singular at $p = \pi$. At singularity, the following finite-size ansatz can be made

$$S_1(\gamma, p = \pi, N) \xrightarrow{N \to \infty} r_1 \frac{\eta_1}{\eta_1 - 1} \left[1 - \sin(\eta_1 \pi / 2)(a_1/N)^{\eta_1 - 1} \right]. \tag{33}$$

Numerical data compared with this ansatz are shown in Fig.[8].

5 AFH model with next-to-nearest neighbour coupling and external magnetic field

Antiferromagnetic order is manifested in the Heisenberg model via singularities of structure factors when $p \to \pi$, $N \to \infty$ and $z = (1 - p/\pi)N = $ constant. Then the transverse structure factor becomes infinite, while the longitudinal structure factor remains finite. This antiferromagnetic order can be destroyed (a) by switching on an external uniform magnetic field B or (b) by frustration (when nearest neighbour and next-to-nearest neighbour coupling do not favour the same alignment pattern). Such a case will be examined in this section. The rersults reveal that there are similarities between the anisotropic chain (XXZ) and the chain with next-to-nearest neighbour coupling.

5.1 Presence of external magnetic field

In the presence of a uniform external magnetic field $\mathbf{B}$, Eq.(4), (with $\mathbf{B}_i = \mathbf{B} = Be_z$ and $J_B = 1$), one observes in the spectrum of the Heisenberg Hamiltonian the following property: Zero-frequency modes $\omega(p_{soft}) = 0$ emerge at non-zero "soft-mode" momenta p_{soft}. These modes show up in the dynamical stucture

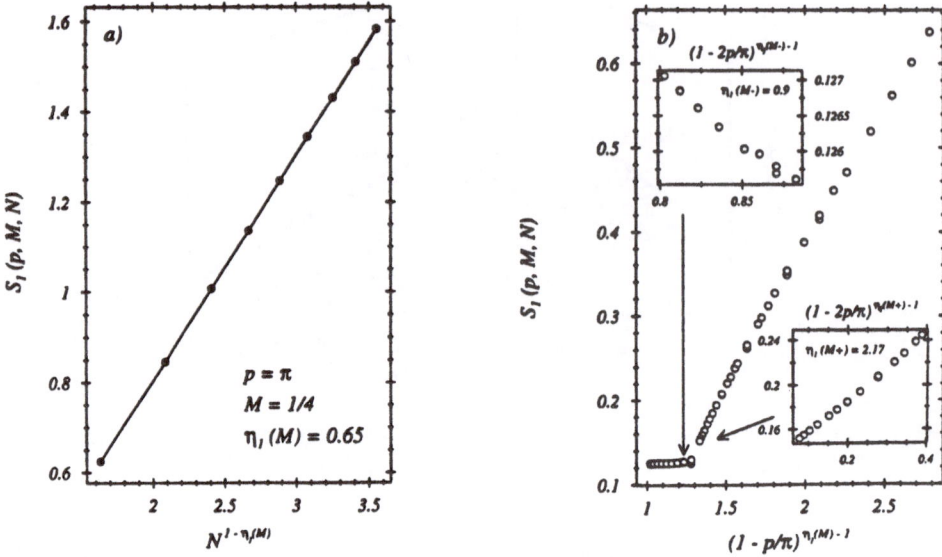

Fig. 10. Transverse structure factor at $M = 1/4$. (a) Finite-size behaviour at $p = \pi$, Eq.(35). (b) Momentum dependence for $p \to \pi$, Eq.(36).

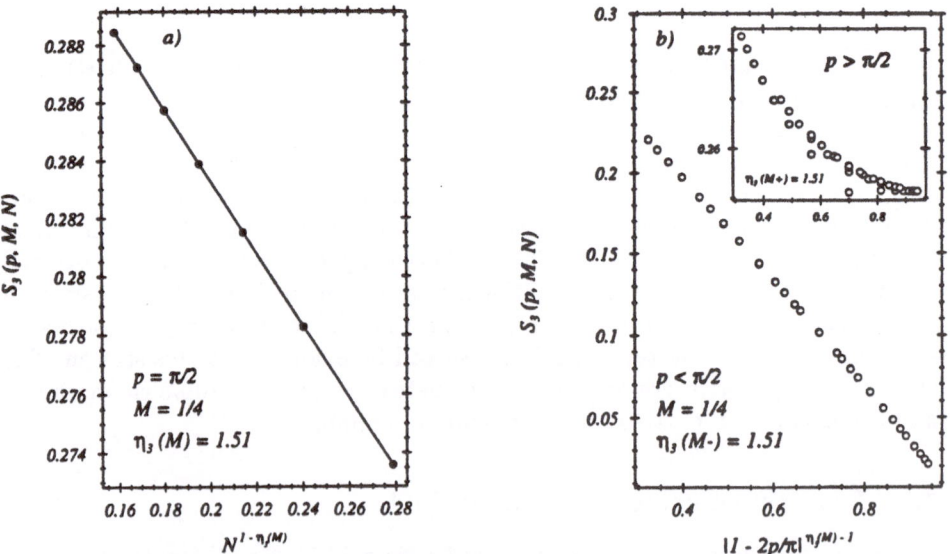

Fig. 11. Longitudinal structure factor at $M = 1/4$. (a) Finite-size behaviour at soft-mode $p_3(M)$, Eq.(38). (b) Momentum dependence for $p \to p_3(M)$, Eq.(39).

factors $S_1(\omega, p)$ and $S_3(\omega, p)$ (Fledderjohann 1996). The momenta depend on the field B, or the magnetization M, respectively. One has

$$p_1(M) = 2\pi M, \quad p_3(M) = \pi(1 - 2M). \tag{34}$$

The critical exponents η_j are supposed to govern the long distance behaviour of the static structure factors and the infra-red behaviour of the dynamical structure factors. Now $\eta_i = \eta_i(M)$ become field dependent. Provided that the low-lying excitation spectrum is governed by conformal symmetry, one can make analytical predictions. Numerical studies (Karbach 1995), (Fledderjohann 1996) show the following behaviour. In Fig.[9] the longitudinal structure factor is plotted against momentum and magnetization. The results correspond to chains of $N = 20, 22, \cdots, 28$ spins. The structure factor is a smooth function except for a ridge at the soft-mode. One observes that the soft-mode moves with magnetization. On the other hand, the transverse structure factor diverges at $p = \pi$ when $N \to \infty$ like

$$S_1(p = \pi, M, N) \longrightarrow_{N \to \infty} N^{1 - \eta_1(M)}. \tag{35}$$

For $M = 1/4$ one finds, e.g., $\eta_1 = 0.65$, shown in Fig.[10a]. The exponent η_1 governs also the approach to the singularity at $p = \pi$,

$$S_1(p, M, N = \infty) \longrightarrow_{p \to \pi} (1 - p/\pi)^{\eta_1(M) - 1}, \tag{36}$$

which is shown in Fig.[10b]. When approaching the soft-mode momentum $p_1(M)$, one finds the behaviour

$$S_1(p, M, N = \infty) \longrightarrow_{p \to p_1(M) \pm 0} (1 - p/p_1(M))^{\eta_1^{\pm}(M) - 1}. \tag{37}$$

Numerical results indicate that $\eta_1^-(M)$ and $\eta_1^+(M)$ do not necessarily agree. The corresponding behaviour of the longitudinal stucture factor is as follows. The finite size behaviour at the soft-mode momentum $p_3(M)$ is given by

$$S_3(p_3(M), M, N) \longrightarrow_{N \to \infty} N^{1 - \eta_3(M)}. \tag{38}$$

This is displayed in Fig.[11a]. Approaching the soft-mode from the left one has

$$S_3(p, M, N = \infty) \longrightarrow_{p \to p_3(M) - 0} (1 - p/p_3(M))^{\eta_3^-(M) - 1}, \tag{39}$$

shown in Fig.[11b]. There is indication for $\eta_3(M) = \eta_3^-(M)$. However, if $\eta_3^+(M)$ and $\eta_3^-(M)$ agree is an open question.

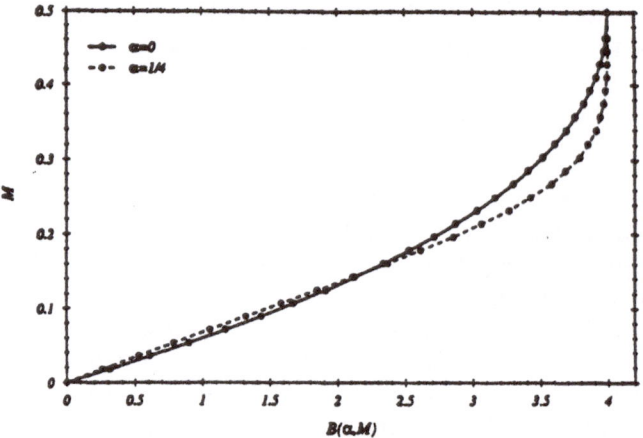

Fig. 12. Magnetization curve for $\alpha = 0$ and $\alpha = 1/4$. The solid line is the Bethe ansatz solution with N=2048 spins.

5.2 Next-to-nearest neighbour coupling

Now let us consider the model Hamiltonian including next-to-nearest neighbour coupling and the presence of a magnetic field, given by a combination of Eqs.(5,4). This model has a rich phase structure, depending on the parameters J_1, J_2 and B. There is a ferromagnetic phase, an antiferromagnetic phase and a frustrated phase (Farnell 1994). Haldane (Haldane 1982) has shown that within the antiferromagnetic phase there is a phase transition at α_c between a spin-fluid phase ($\alpha < \alpha_c$) and a dimer phase ($\alpha > \alpha_c$). The transition point has been determined to be $\alpha_c = 0.24$ (Okamoto 1992). Related to this is Gluzman's (Gluzman 1994) prediction of the occurence of a plateau in the magnetization curve. Moreover, it is remarkable that the model is exactly soluble at $\alpha = 1/2$, $B = 0$, called Majumdar-Ghosh point (Majumdar 1969). One can ask: Where are the observable differences in the model with and without next-to-nearest neighbour coupling, i.e., $\alpha \neq 0$ and $\alpha = 0$?

Numerical studies by Schmidt et al. (Schmidt 1996) and Gerhardt et al. (Gerhardt 1996) give the following information. Firstly, one observes in the magnetization curve near saturation the following characteristic behaviour,

$$M(\alpha = 1/4, B) \longrightarrow_{B \to B_s} 1/2 - \text{const.} \times (B_s - B)^{1/4}, \qquad (40)$$

in contrast to the square root behaviour for $\alpha = 0$, given by Eq.(13). This is shown in Fig.[12]. In the presence of an external magnetic field and next-to-nearest neighbour coupling, the structure factors develop singularities at soft-mode momenta, given by Eq.(34). Now the strength of the singularity will depend on the frustration parameter α. The transverse structure factor and the longitudinal structure factor for fixed values of α but different values of magnetization

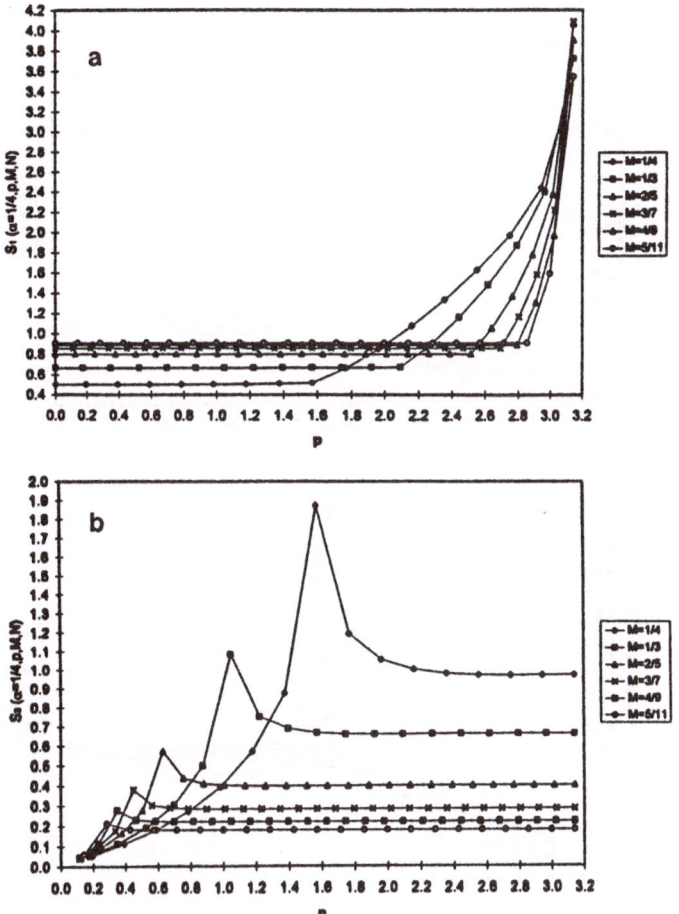

Fig. 13. Transverse structure factor (a) and longitudinal structure factor (b) versus momentum p for fixed magnetization $M = 1/4$ but different values of α.

M are shown in Fig.[13a] and [13b], respectively. Fig.[14] displays the structure factors for fixed M but different values of α. The critical exponents η_i will now depend on M and α. The finite-size dependence of the transverse structure factor at $p = \pi$ is described by

$$S_1(\alpha, p = \pi, M, N) \longrightarrow_{N\to\infty} A_1(\alpha, M) + B_1(\alpha, M)N^{1-\eta_1(\alpha,M)}. \quad (41)$$

E.g., one finds $\eta_1(\alpha = 0, M = 1/4) = 0.65$, while $\eta_1(\alpha = 1/4, M = 1/4) = 1.16$. Similarly, the longitudinal structure factor at the soft-mode $p_3(M)$ behaves as

$$S_3(\alpha, p = p_3(M), M, N) \longrightarrow_{N\to\infty} A_3(\alpha, M) + B_3(\alpha, M)N^{1-\eta_3(\alpha,M)}. \quad (42)$$

E.g., one finds $\eta_3(\alpha = 0, M = 1/4) = 1.5$, while $\eta_3(\alpha = 1/4, M = 1/4) = 0.84$. Those critical exponents extracted from Eqs.(41,42) are given in Tab.[1] as a

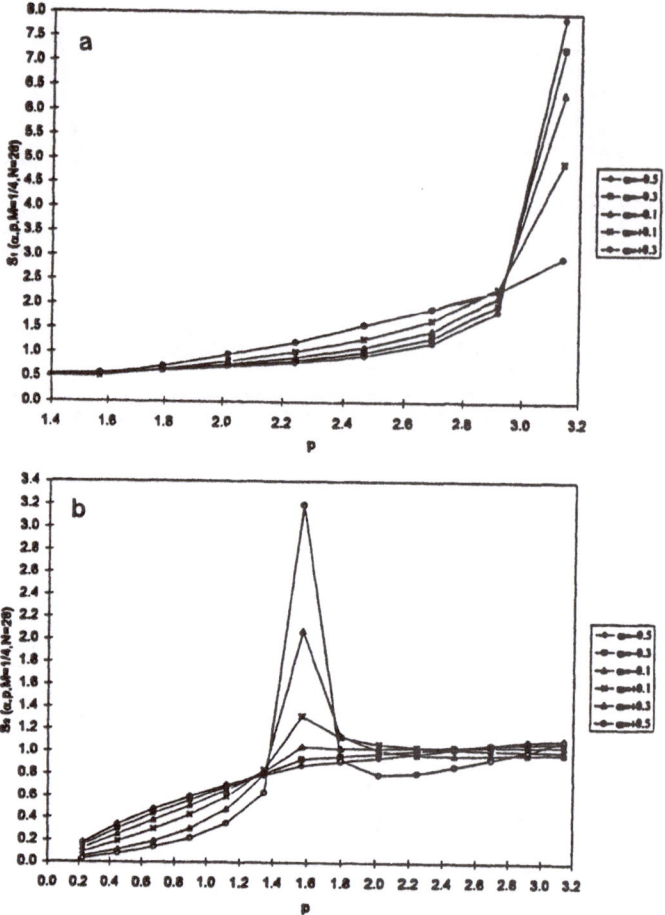

Fig. 14. Transverse structure factor (a) and longitudinal structure factor (b) versus momentum p for fixed $\alpha = 1/4$ but different values of M.

function of α. Based on the assumption that conformal symmetry holds, one would expect $\eta_1 \cdot \eta_3 = 1$. The numerical data for most values of α deviate from the expected result by a few percent. However, special attention should be paid to the region $\alpha \sim -0.2$ and the region $\alpha \geq 0.4$. For $\alpha = -0.2$, Eq.(42) does not yield a value of η_3. Our results show that the finite-size dependence is very weak in this region ($B_3(\alpha, M)$ is very small)) and may even vanish for some value of α. Secondly, for $\alpha \geq 0.4$, $\eta_1 \cdot \eta_3$ deviates substantially from the value one. The reason is most likely the following: The structure factors have been computed in the sector corresponding to quantum numbers of the Heisenberg model, Eq.(9). For $\alpha \geq 0.4$, those are presumably no longer the quantum numbers of the ground state, hence the critical exponents η_i do not correspond to the ground state but to some excited state.

Table 1. Critical exponents η_i for $M = 1/4$ versus α

α	-0.5	-0.4	-0.3	-0.2	-0.1	0	0.1	0.2	0.25	0.3	0.4	0.5
η_1	0.39	0.42	0.45	0.50	0.56	0.65	0.78	1.00	1.16	1.36	2.74	3.53
η_3	2.52	2.45	2.34	?	1.66	1.49	1.26	0.98	0.84	0.70	0.47	0.32
$\eta_1 \cdot \eta_3$	0.98	1.03	1.06	?	0.93	0.97	0.98	0.98	0.97	0.94	1.28	1.13

6 Summary and outlook

The 1-D spin 1/2 AFH model in the presence of an external magnetic field and with next-to-nearest neighbour coupling can be solved "exactly" by diagonalization (via Lançzos) of chains of up to 50 spins (depending on M). The ground state properties of the model yield data with high numerical precision allowing for finite-size scaling analysis and extraction of critical exponents. Here we have concentrated on static spin structure factors. One finds a different behaviour for $\alpha < \alpha_c$ and $\alpha > \alpha_c$. The difference is observed in the strength of the singularities of structure factors and in the magnetization curve near saturation. Much interesting physics is to be explored when going into D=2,3 dimensions (e.g., prediction of plateaus in the magnetization curve). Exact diagonalization is presently limited in D=2 to 7×7 spins (with helical boundary conditions). To analyze critical behaviour in D=2, new methods yielding numerically precise results for a larger number of spins are needed.

Acknowledgement

One of the authors (H.K.) is grateful for support by NSERC Canada.

References

I. Affleck, D. Gepner, H. Schulz and T. Ziman, J. Phys. A22(1989)511.

H.M. Babujian, Nucl. Phys. B215(1983)317.

R. J. Baxter, Exactly Solved Models in Statistical Mechanics, Academic Press (1980).

J. Bernasconi, T. Schneider, Physics in One Dimension, Springer, New York (1981).

H.A. Bethe, Z. Phys. 71(1931)205.

R.F. Bishop, in Recent Progress in Many-Body Theories, eds. E. Schachinger et al., Plenum, New York (1995), p.195;
see also R.F. Bishop, talk at this Workshop.

N.M. Bogoliubov, A.G. Izergin and V.E. Korepin, Nucl. Phys.B275(1986)687.

J.C. Bonner and M.E. Fisher, Phys. Rev. 135(1964)17640.

J.L. Cardy, Nucl. Phys. B270(1986)186.

J.L. Cardy, in Phase Transitions and Critical Phenomena, Ed. C. Domb and J.L. Lebowitz, Academic Press (1987), Vol.11.

J.L. Cardy, ed., Finite-Size Scaling, North Holland, Amsterdam (1988).

D.M. Ceperley, in Recent Progress in Many-Body Theories, eds. E. Schachinger et al., Plenum, New York (1995), p.455.

D.C. Dender, D. Davidović, D.H. Reich, C. Broholm, K. Lefmann and G. Aeppli, Phys. Rev. B53(1996)2583.

J.D. Dunitz, Acta Crist. 10(1957)307.

F.H.L.Essler, H. Frahm, A.R. Its and V.E. Korepin, solv-int/9604005.

K. Fabricius, K.H. Mütter and H. Grosse, Phys. Rev. B42(1990)4656.

K. Fabricius, U. Löw, K.H. Mütter and P. Ueberholz, Phys. Rev. B44(1991)7476.

D.J.J. Farnell and J.B. Parkinson, J. Phys. Condens. Matter 6(1994)5521.

M.E. Fisher, in Critical Phenomena, Proceed. 51st Enrico Fermi Summer School, ed. M.S. Green, Academic Press, London (1971).

M.N. Fisher and M.N. Barber, Phys. Rev. Lett. 28(1972)1516.

A. Fledderjohann, C. Gerhardt, K.H. Mütter, A. Schmitt and M. Karbach, Phys. Rev. B54(1996)7168.

C. Gerhardt, A. Fledderjohann, E. Aysal, K.H. Mütter, J.F. Audet and H. Kröger, J. Phys. Condens. Matter, subm.

T. Giamarchi and H.J. Schulz, Phys. Rev. B39(1989)4620.

S. Gluzman, Phys. Rev. B49(1994)11962.

R.B. Griffith, Phys. Rev. A133(1964)768.

Z.N.C. Ha, Phys. Rev. Lett. 73(1994)1574.

F.D.M. Haldane, Phys. Rev. B25(1982)4928; B26(1982)5257.

F.D.M. Haldane, Phys. Rev. Lett. 60(1988)635.

K.A. Hallberg, P. Horsch and G. Martinez, Phys. Rev. B52(1995)R719.

K. Hallberg, X.Q.G. Wang, P. Horsch and A. Moreo, Phys. Rev. Lett. 76(1996)4955.

L. Hulthén, Ark. Mat. Astr. Fys. 26(1938)1.

M. Inui, S. Doniach and M. Gabay, Phys. Rev. B38(1988)6631.

J.D. Johnson, S. Krinsky and B.M. McCoy, Phys. Rev. A8(1973)2526.

M. Karbach and K.H. Mütter, Z. Phys. B90(1993)83.

M. Karbach, Ph.D. Thesis, Universität Wuppertal (1994a).

M. Karbach, K.H. Mütter and M. Schmidt, Phys. Rev. B50(1994b)9281.

M. Karbach, K.H. Mütter and M. Schmidt, J. Phys. Condens. Matter. 7(1995)2829.

K.J.B. Lee and P. Schlottmann, Phys. Rev. B36(1987)466.

M. Lüscher, Nucl. Phys. B117(1976)475.

A. Luther and I. Peschel, Phys. Rev. B12(1975)3908.

C.K. Majumdar and D.K. Ghosh, J. Math. Phys. 10(1969)1388; 1399.

F. Mazzi, Acta Cryst. 8(1955)137.

N.D. Mermin and H. Wagner, Phys. Rev. Lett. 17(1966)1133.

G. Müller, H. Thomas and J.C. Bonner, Phys. Rev. B24(1981)1429; J. Phys. C14(1981)3399.

K. Okamoto and K.Nomura, Phys. Lett. A169(1992)433.

R.A.Römer and B. Sutherland, Phys. Rev. B48(1993)6058.

M. Schmidt, C. Gerhardt, K.H. Mütter and M. Karbach, J. Phys. Condens. Matter 8(1996)553.

B.S. Shastry, Phys. Rev. Lett. 60(1988)639.

R.R.P. Singh, M.E. Fisher and R. Shankar, Phys. Rev. B39(1989)2562.

J. Suzuki, T. Nagao and M. Wadati, Int. J. Mod. Phys. B6(1992)1119.

D.A. Tennant, R.A. Cowley, S.E. Nagler and A. Tsvelik, Phys. Rev. B52(1995)13368.

S.R. White, Phys. Rev. B48(1993)10345.

C.N. Yang and C.P. Yang, Phys. Rev. 150(1966)321; ibid 150(1966)327; ibid 151(1966)258.

M.R.Zirnbauer and F.D.M. Haldane, cond-mat/9504108.

The Z(2) Lattice Gauge Vacuum and the Transverse Ising Model: Two Sides of a Coin

M. L. Ristig[1], J. W. Kim[1], and J. W. Clark[2]

[1] Institut für Theoretische Physik
Universität zu Köln, D-50937 Köln, Germany

[2] McDonnell Center for the Space Sciences
and Department of Physics,
Washington University, St. Louis, MO 63130

Abstract. The uncharged sector of the $Z(2)$ lattice gauge model in two spatial dimensions is analyzed within variational correlated basis functions (CBF) theory applied to the dual Ising-spin model with a transverse magnetic field. The CBF analysis of the associated ground and excited states is conducted with correlated trial wave functions of the Hartree-Jastrow type for Pauli spins on an infinitely extended square lattice with unit lattice constant. The hypernetted-chain (HNC) formalism is utilized for evaluation of the spatial spin-distribution function, the spin-exchange strength, and the ferromagnetic order parameter corresponding to a given trial function of this class. The optimal distribution function and optimal order parameter which minimize the ground-state energy in the thermodynamic limit are determined by solving a pair of Euler-Lagrange equations. The CBF treatment becomes exact in both weak-coupling and strong-coupling limits of the model. We report, in HNC approximation, results for various gross quantities including the ground-state energy, the potential component, the optimal magnetizations, and the optimal magnon energies, as functions of the coupling parameter $0 \leq \lambda \leq \infty$ measuring the strength of the transverse field. We further discuss numerical results for the optimized spin-correlation function, the corresponding static structure function, and the correlation length. The continuous phase transition from the ordered to the disordered phase occurs at a theoretical value $\lambda = \lambda_c \simeq 3.14$ for the coupling. This value is in excellent agreement with earlier results derived from perturbation expansions in conjunction with Padé techniques. At the critical point the ferromagnon energy gap vanishes and the correlation length of the optimal distribution function diverges. The Euler-Lagrange equations admit a metastable region of disorder at coupling constants $\lambda < \lambda_c$.

1 Introduction

Theoretical command of the correlated structure and the phase transitions of crystalline materials (Binder 1991), (Lüthi and Rehwald 1981) will play a key role in modern efforts in materials science aimed at the synthesis of novel materials. The structure of these new materials will in general be quite complex, and the operative interaction mechanisms can be subtle. It will often be necessary to invoke the sophisticated group theory of point and space groups to accounting for the underlying symmetries (Gernoth 1997). Moreover, an adequate theoretical approach must transcend standard mean-field treatments, since interest is focused, experimentally and theoretically, on the crystal behavior at intermediate ranges of coupling strength not far from the phase-transition regions (Ristig and Kim 1996).

To advance our quantitative understanding of the structure and phase transitions of complex crystalline materials, it is advisable to build upon the vast body of theoretical knowledge on pseudospin-lattice systems. The existing literature furnishes useful models for insulating magnetic systems and other relatively simple materials with well understood physical aspects. Among a number of examples, we may point to hydrogen-bonded ferroelectrics, such as KH_2PO_4, where protons move from one minimum of a double well potential to the other. Pseudospin-lattice models offer rich ground for the application of group-theoretic tools to the analysis of symmetries and for the systematic development of correlation effects.

As distinct from stochastic computational approaches and other simulation methods, quantum many-body theory provides powerful analytic or semi-analytic tools to bear on fundamental problems of physical and chemical interest. The coupled cluster (CC) method (Bishop, Kendall, Wong, and Xian 1993) and the correlated basis functions (CBF) approach (Clark and Feenberg 1959) are by now generally regarded as two most successful microscopic formulations currently available. In a large number of applications they have demonstrated their versatility, their ability to achieve very high accuracy at attainable levels of approximation, and their capacity for systematic improvements. CC theory is widely adopted as the method of choice in quantum chemistry, while CBF theory surpasses all other semi-analytic approaches in the microscopic description of quantum fluids and solids in various isotropic and spatial configurations (Campbell 1978), (Gernoth, Clark, Senger, and Ristig 1994), (Gernoth, Clark, and Ristig 1995), (Clements et. al. 1994).

The CC and CBF approaches have recently been employed in studies of the uncharged sector of the $U(1)$ lattice gauge model in two spatial dimensions (Bishop, Kendall, Wong, and Xian 1993), (Bishop, Kendall, Wong, and Xian 1993), (Bishop 1995), (Bishop, Davidson, and Xian 1995), (Dabringhaus, Ristig, and Clark 1991), (Dabringhaus and Ristig 1991). In addition, the CC method has been applied for an analysis of the vacuum properties of the $Z(2)$ lattice gauge model(Bishop, Kendall, Wong, and Xian 1993) and spin lattice models (Bishop, Parkinson, and Xian 1992).

The Ising model in a transverse magnetic field on a simple hypercubic lattice is a prototypical spin-lattice system that has received much attention over many decades (Rushbrooke et. al. 1974). This model may be used to test the quality of approximations based on CBF and other many-body theories, before proceeding to more complex lattice systems or other crystalline materials. In further steps, one may develop analogous microscopic descriptions of other spin-lattice models (Bishop, Parkinson, and Xian 1992) such as the XY model or lattices with more complex space symmetries (Gernoth 1997). In two spatial dimensions ($D = 2$), the transverse Ising model is defined by a system of N spins on the lattice sites of an infinitely extended simple quadratic lattice. The spins interact pairwise via a potential of the Ising-type and are subject to a transverse external field.

In the present work, this classic problem will be treated within the framework of variational-CBF theory. An intriguing feature of the $D = 2$ transverse Ising model is that it is dual to the Z(2) lattice gauge model specified on a square lattice in two spatial dimensions (Kramers and Wannier 1941), (Savit 1980). Duality implies that the two systems possess the same excitation spectrum and the same critical behavior. A successful microscopic description of one system yields a corresponding microscopic understanding of the other. The Z(n) models are especially interesting for gauge theory because they exhibit confinement and transitions between different phases. In the limit $n \to \infty$ they can further be related to compact quantum electrodynamic U(1) theory.

The Z(2) system in question exhibits two phases, an ordered phase in the strong-coupling region and a disordered phase in the week-coupling regime. The system is confined in the ordered phase, where the magnetic monopoles are condensed and the elementary excitations are glueballs. In the disordered phase, the low-lying excitations are magnetic monopoles with non-zero mass (Horn, Weinstein, and Yankielowicz 1979) In the dual picture – i.e., on the spin-lattice side of the coin – the ordered phase is ferromagnetic, with ferromagnon excitations that become soft at the phase boundary, while the disordered phase is paramagnetic. The paramagnon excitations, counterparts of the magnetic monopoles of the Z(2) model, have nonzero energy gap.

The $Z(2)$ model and its dual spin system have been extensively studied and have served as a testing ground for a number of analytic methods and computational techniques, including perturbation-theoretic expansions and Padé approximations (Pfeuty and Elliott 1971), (Stinchcombe 1973), (Hamer and Irving 1984), real-space renormalization-group approaches (Penson, Jullien, and Pfeuty 1979), (Fradkin and Raby 1979), variational descriptions with parametrized Ansätze (Dagotto and Moreo 1984), (Dagotto et. al. 1985), finite lattice techniques (Roomany and Wyld 1980), stochastic procedures (Creutz, Jacobs, and Rebbi 1983), (Allton, Yung, and Hamer 1989), and others treatments (Di Bartolo, Gambini, and Trias 1989). The results derived from CC theory are reported in Refs.(Bishop, Kendall, Wong, and Xian 1993), (Bishop 1995). The various studies provide numerical results or estimates for the critical coupling parameter where the order-disorder transition occurs, for the ground-state energies, and for the energies of the vacuum excitations.

Sec. II provides the necessary foundation for performing a complete variational CBF analysis. Sec. III presents the hypernetted-chain (HNC) equations that permit explicit evaluation of the energy functional. The Euler-Lagrange equation that determines the optimal spatial correlation function is analyzed in Sec. IV and is cast into the form of a paired-magnon equation. The associated Feenberg effective potential is constructed, and the HNC equations for calculating the required generalized distribution function are given. Sec. V deals with the renormalized Hartree equation for the optimal magnetization. Our numerical results are reported and discussed in Sec. VI.

2 Basic Relations

The correlated-wave-function or variational-CBF analysis of the ground and excited states of the two-phase $Z(2)$ system is carried out within the dual framework of a system of Ising spins fixed on the sites of an infinitely extended quadratic lattice and interacting with a transverse external field. The energetics and dynamical behavior of the spin system are characterized by the Hamiltonian

$$\mathcal{H} = \frac{1}{2} \sum_{i,j}^{N} \Delta_{ij} \sigma_i^x \sigma_j^x + \lambda \sum_i^N (1 - \sigma_i^z) \quad , \tag{1}$$

for N spins (or N lattice sites) in the (x, y) plane, with $N \to \infty$. The Pauli-spin operator σ_i at lattice point $\mathbf{r}_i$ has the components σ_i^x and σ_i^z in the x- and z-directions, with eigenvalues ± 1. The strength of the external field is measured by the coupling parameter λ $(0 \le \lambda \le \infty)$. The spin-spin interaction is of the Ising type,

$$\Delta(\mathbf{n}) = \begin{cases} 2D & , \quad \mathbf{n} = 0 \quad , \\ -1 & , \quad \text{for nearest neighbors} \quad , \\ 0 & , \quad \text{otherwise} \quad , \end{cases} \tag{2}$$

with $\Delta_{ij} \equiv \Delta(\mathbf{r}_i - \mathbf{r}_j) = \Delta(\mathbf{n})$, while the spatial dimension is $D = 2$. We note that the formalism employed here may be applied to spin models in higher dimensions D and having more general Ising interactions $\Delta(\mathbf{n})$ that extend beyond nearest neighbors.

A microscopic analysis of the ground and excited states corresponding to the Hamiltonian (1) requires the evaluation of (i) the magnetization

$$M = \frac{\langle \Psi | \sigma_i^x | \Psi \rangle}{\langle \Psi | \Psi \rangle} \quad , \tag{3}$$

which serves as an order parameter for the ordered (ferromagnetic) phase, (ii) the transverse magnetization

$$A = \frac{\langle \Psi | \sigma_i^z | \Psi \rangle}{\langle \Psi | \Psi \rangle} \quad , \tag{4}$$

which provides a measure for the magnetic energy contained in the system, and (iii) the spatial correlation function

$$g(\mathbf{n}) = \frac{\langle \Psi | \sigma_i^x \sigma_j^x | \Psi \rangle}{\langle \Psi | \Psi \rangle} \quad , \tag{5}$$

with $\mathbf{n} = \mathbf{r}_i - \mathbf{r}_j$. The expectation values may be taken with respect to the true many-body ground state or to a suitably chosen set of correlated trial ground states. Quantities (3)-(5) determine the expectation value $E = \langle \Psi | \mathcal{H} | \Psi \rangle / \langle \Psi | \Psi \rangle$

of the ground-state energy and the energy of the elementary excitations, in Feynman approximation (Feynman 1954), (Feenberg 1969).

The present analysis employs a set of correlated many-body wave functions of Hartree-Jastrow type (Ristig and Kim 1996),

$$|\Psi\rangle = \exp(MU_M + U)|0\rangle \quad , \tag{6}$$

where the vacuum reference state $|0\rangle$ is taken as a symmetric product of single-spin eigenstates of the spin operators σ_i^z, $i = 1, \cdots, N$, all with eigenvalue $+1$. The pseudopotentials defining the exponential correlating factor are given by

$$U = \frac{1}{2} \sum_{i<j}^{N} u(\mathbf{r}_{ij}) \sigma_i^x \sigma_j^x \tag{7}$$

and

$$U_M = \sum_i^N u_1(\mathbf{r}_i) \sigma_i^x + \frac{1}{4} \sum_{i<j}^{N} u_M(\mathbf{r}_{ij})(\sigma_i^x + \sigma_j^x) \quad , \tag{8}$$

where the functions $u_1(\mathbf{r}_i)$, $u(\mathbf{r}_{ij})$, and $u_M(\mathbf{r}_{ij})$ should be determined to minimize the ground-state energy expectation value. For an infinitely extended lattice (thus $N \to \infty$), the function $u_i(\mathbf{r}_i)$ is constant and independent of the lattice site, while $u(\mathbf{r}_{ij})$ and $u_M(\mathbf{r}_{ij})$ depend only on the relative distance $|\mathbf{n}| = |\mathbf{r}_i - \mathbf{r}_j|$.

The energy expectation value associated with ansatz (6),

$$E = \frac{\langle \Psi | \mathcal{H} | \Psi \rangle}{\langle \Psi | \Psi \rangle} = V + V_M \quad , \tag{9}$$

has the potential energy component

$$V = \frac{1}{2} \sum_{i,j}^{N} \Delta_{ij} \frac{\langle \Psi | \sigma_i^x \sigma_j^x | \Psi \rangle}{\langle \Psi | \Psi \rangle} \tag{10}$$

and the magnetic energy part

$$V_M = \lambda \sum_i^N \frac{\langle \Psi | (1 - \sigma_i^z) | \Psi \rangle}{\langle \Psi | \Psi \rangle} \quad . \tag{11}$$

Following Ristig and Kim (Ristig and Kim 1996), we express the thermodynamic ($N \to \infty$) limit of energies (10) and (11), per spin or lattice site, as

$$\frac{V}{N} = D(1 - M^2) \left[1 + \frac{1}{2D} \sum_{\mathbf{n}} \Delta(\mathbf{n}) G(\mathbf{n}) \right] \quad , \tag{12}$$

$$\frac{V_M}{N} = \lambda \left[1 - (1 - M^2)^{\frac{1}{2}} n_{12} \right] \quad . \tag{13}$$

The relations (12) and (13) involve the order parameter (3) (magnetization in the x direction) and the spin-exchange strength

$$n_{12} = (1 - M^2)^{-\frac{1}{2}} A \quad , \tag{14}$$

which in turn involves the transverse magnetization (4).

To separate the kinetic molecular-field effects from the dynamic correlation effects it is convenient to work with the modified distribution function $G(\mathbf{n})$ defined by the decomposition of quantity (5),

$$g(\mathbf{n}) = \delta_{\mathbf{n},0} + (1 - \delta_{\mathbf{n},0})M^2 + (1 - M^2)G(\mathbf{n}) \quad , \tag{15}$$

where $G(\mathbf{n})$ is the short-range part of the spatial correlations with $G(\mathbf{n}) = 0$ for $\mathbf{n} = 0$ and $\mathbf{n} \to \infty$.

It is to be noted that the exchange factor (14) appearing in expression (13) depends on the distribution function $G(\mathbf{n})$ and, in the ferromagnetic phase, on the order parameter (3). We must therefore consider the energy per spin (9) as a functional $E/N = e[G(\mathbf{n}), M, \lambda]$ in the thermodynamic limit, where $N \to \infty$ at fixed lattice constant. In a mean-field approximation, correlation effects are ignored and one has $G(\mathbf{n}) \equiv 0$ and $n_{12} \equiv 1$. Consequently the energy functional (9) per lattice site reduces to a simple function of the quantities M and λ,

$$\frac{E}{N} = (1 - M^2)D + \lambda\big[1 - (1 - M^2)^{\frac{1}{2}}\big] \quad , \tag{16}$$

yielding $E/N = D$ in the disordered phase and

$$\frac{E}{N} = \lambda\Big(1 - \frac{\lambda}{4D}\Big) \quad , \qquad 0 \le \lambda \le 2D \quad , \tag{17}$$

for the optimal order parameter $M = \big[1 - (\lambda/2D)^2\big]^{\frac{1}{2}}$ that minimizes the energy (9) in the ordered phase. CBF theory takes account of spatial correlations and provides the analytic tools for an explicit construction of the more complex energy functional (9) in terms of the quantities $G(\mathbf{n})$, M and λ.

We next construct the full set of correlated many-body wave functions. In analogy with the density fluctuation operators employed in the CBF theory of quantum fluids (Feenberg 1969) such as liquid ^{4}He we introduce the excitation operators $\rho_{\mathbf{k}}^x = \sum_i^N e^{i\mathbf{k}\cdot\mathbf{r}_i}\sigma_i^x$ and form the ideal magnon states

$$|\Psi_{\mathbf{k}}\rangle = \rho_{\mathbf{k}}^x|\Psi\rangle \quad . \tag{18}$$

In Feynman approximation (Feynman 1954), the energies of the low-lying elementary excitations are given by

$$\omega(\mathbf{k}) = \frac{\epsilon}{S(\mathbf{k})} \quad , \tag{19}$$

in terms of the renormalized Hartree quasi-particle energy (Ristig and Kim 1996)

$$\epsilon = \frac{\langle\Psi|[\rho_{\mathbf{k}}^x, [\mathcal{H}, \rho_{-\mathbf{k}}^x]]|\Psi\rangle}{\langle\Psi|\Psi\rangle} = 2\lambda n_{12}(1 - M^2)^{-1/2} \quad , \tag{20}$$

and the static structure function

$$S(\mathbf{k}) = 1 + \sum_{\mathbf{n}} e^{i\mathbf{k}\cdot\mathbf{n}} G(\mathbf{n}) \quad . \tag{21}$$

The corresponding expectation values (3)-(5) may be evaluated by applying hypernetted-chain (HNC) techniques (Feenberg 1969), (Clark 1979), (Ristig, Fantoni, and Kürten 1983), (Ristig, Kim, and Mehlmann 1996), (Ristig 1981).

3 Hypernetted-Chain Equations

The short-ranged spatial distribution function $G(\mathbf{n})$ appearing in expression (15) may be decomposed into a nodal component $N(\mathbf{n})$ and a non-nodal part

$$G(\mathbf{n}) = X(\mathbf{n}) + N(\mathbf{n}) \quad . \tag{22}$$

These components are related by a chain equation

$$N(\mathbf{n}) = \sum_{\mathbf{m}} X(\mathbf{n} - \mathbf{m})[X(\mathbf{n}) + N(\mathbf{n})] \quad , \tag{23}$$

which holds at any value of the order parameter M. A second relation between the quantities $X(\mathbf{n})$ and $N(\mathbf{n})$ is provided by a hypernet equation of the form (Ristig and Kim 1996)

$$X(\mathbf{n}) = 1 - N(\mathbf{n}) - 2\Big\{1 + [1 + (1 - M^2)(\exp 4C(\mathbf{n}) - 1)]^{1/2}\Big\}^{-1} \tag{24}$$

at $\mathbf{n} \neq 0$. The exponential is generated by the quantity

$$C(\mathbf{n}) = u(\mathbf{n}) + (1 - M^2)^{-1} N(\mathbf{n}) + E(\mathbf{n}) \quad , \tag{25}$$

where $E(\mathbf{n})$ represents the elementary contributions. For $\mathbf{n} = 0$, Eq. (24) reduces to the simple relation $X(0) = 1 - N(0)$. In the HNC/0 approximation, the elementary-diagram sum $E(\mathbf{n})$ is set equal to zero. Thereupon, Eqs. (23) and (24) become a closed set of relations for the functions $X(\mathbf{n})$ and $N(\mathbf{n})$, under the proviso that the generator $u(\mathbf{n})$ is available as input at a given magnetization M.

The hypernetted-chain analysis of the exchange strength (14) is based on the exponential representation (Ristig and Kim 1996)

$$n_{12} = \exp\Big[D_{12} - \frac{1}{2}(D_1 + D_2)\Big] \quad . \tag{26}$$

The functionals D_1, D_2, and D_{12} may be expressed in terms of functions $X(\mathbf{n})$, $N(\mathbf{n})$ and an additional set of modified nodal and non-nodal quantities, $\hat{X}(\mathbf{n})$ and $\hat{N}(\mathbf{n})$. Employing these functions, the exponent (26) reads explicitly

$$\ln n_{12} = \frac{1}{2}\sum_{\mathbf{n}} \ln[1 - G(\mathbf{n})] + \frac{1}{4}\sum_{\mathbf{n}} \ln\Big[1 + \frac{1 - M}{1 + M}G(\mathbf{n})\Big]$$

$$+ \frac{1}{4}\sum_{\mathbf{n}} \ln\left[1 + \frac{1+M}{1-M}G(\mathbf{n})\right] + \frac{1}{2}(1-M^2)^{-1}\sum_{\mathbf{n}} G(\mathbf{n})N(\mathbf{n})$$

$$- \frac{1}{2}\sum_{\mathbf{n}} \ln[1 - M(1+M)\hat{G}(\mathbf{n})] - \frac{1}{2}\sum_{\mathbf{n}} \ln[1 + M(1-M)\hat{G}(\mathbf{n})]$$

$$- M^2 \sum_{\mathbf{n}} \hat{N}(\mathbf{n}) + \frac{1}{2}M^2 \sum_{\mathbf{n}} G(\mathbf{n})\hat{N}(\mathbf{n})$$

$$- \frac{1}{2}M^2(1-M^2)\sum_{\mathbf{n}} \hat{G}(\mathbf{n})\hat{N}(\mathbf{n}) + E_{12} \quad , \tag{27}$$

with $G(\mathbf{n})$ given by the decomposition (22) and $\hat{G}(\mathbf{n})$ by the analogous relation $\hat{G}(\mathbf{n}) = \hat{X}(\mathbf{n}) + \hat{N}(\mathbf{n})$. The elementary component E_{12} is ignored in HNC/0 approximation.

The HNC equations associated with the modified distribution function $\hat{G}(\mathbf{n})$ take the form (Ristig and Kim 1996)

$$\hat{N}(\mathbf{n}) = (1-M^2)^{-1}\sum_{\mathbf{m}} G(\mathbf{n}-\mathbf{m})G(\mathbf{m}) + \sum_{\mathbf{m}} G(\mathbf{n}-\mathbf{m})\hat{X}(\mathbf{m}) \quad , \tag{28}$$

$$\hat{X}(\mathbf{n}) = -\hat{N}(\mathbf{n}) + (1-\delta_{\mathbf{no}})M^{-1}\frac{\tanh M[\hat{u}(\mathbf{n}) + \hat{N}(\mathbf{n}) + \hat{E}(\mathbf{n})]}{1 + M\tanh M[\hat{u}(\mathbf{n}) + \hat{N}(\mathbf{n}) + \hat{E}(\mathbf{n})]} \quad . \tag{29}$$

The generating pseudopotential $\hat{u}(\mathbf{n})$ appearing in the hypernet equation (29) is defined by

$$\hat{u}(\mathbf{n}) = \frac{1}{4}M^{-1}\left\{\ln\left[1 + \frac{1-M}{1+M}G(\mathbf{n})\right] - \ln\left[1 + \frac{1+M}{1-M}G(\mathbf{n})\right]\right\} \quad . \tag{30}$$

The solutions of the HNC equations (23), (24) and (28), (29) permit the evaluation of the exchange strength (26). In the disordered phase where $M \equiv 0$, the functional D_{12} vanishes and quantities D_1 and D_2 are given (in HNC/0 approximation) by the specialized expression

$$D_1 = D_2 = -\frac{1}{2}\sum_{\mathbf{n}} \ln[1 - G^2(\mathbf{n})] - \frac{1}{2}\sum_{\mathbf{n}} N(\mathbf{n})G(\mathbf{n}) + E_D \quad . \tag{31}$$

In this case, solutions of the HNC equations for the modified functions $\hat{X}(\mathbf{n})$ and $\hat{N}(\mathbf{n})$ is not necessary.

In a second step within CBF theory (Clark and Feenberg 1959), (Feenberg 1969), (Clark 1979), one minimizes the ground-state energy and determines the optimal many-body wave function of the chosen trial set. This process is described in the next section.

4 Paired-Magnon Equation

To optimize the ground-state correlations, i.e., the generator function $u(\mathbf{n})$, we employ the minimum principle for the ground-state energy. The energy is varied with respect to the generator to derive an Euler-Lagrange equation

$$\frac{\delta E}{\delta u(\mathbf{n})} = 0 \qquad (32)$$

for the optimal spatial distribution function $G(\mathbf{n})$, at constant coupling parameter λ and constant magnetization M. This equation can be given the more explicit form (Ristig and Kim 1996)

$$\dot{G}(\mathbf{n}) + \frac{1}{2}\varepsilon(1 - M^2)G(\mathbf{n}) = 0 \quad , \qquad (33)$$

which is the analog of the paired-phonon equation familiar in the CBF theory of quantum fluids(Feenberg 1969). Since the excitations corresponding to the Hamiltonian (1) are magnons, we may interpret Eq. (33) as a paired-magnon equation. The function $\dot{G}(\mathbf{n})$ appearing in Eq. (33) is defined by the derivative

$$\dot{G}(\mathbf{n}) = \sum_{\mathbf{m}} v(\mathbf{m}) \frac{\delta G(\mathbf{n})}{\delta u(\mathbf{m})} \quad , \qquad (34)$$

which involves the Feenberg effective potential(Feenberg 1969)

$$v(\mathbf{n}) = v^*(\mathbf{n}) + M^2 v_M^*(\mathbf{n}) \quad . \qquad (35)$$

The second term vanishes in the disordered state.

Explicit expressions for the components $v^*(\mathbf{n})$ and $v_M^*(\mathbf{n})$ may be derived by varying the energy functional with respect to the distribution function $G(\mathbf{n})$. To this end we employ the expression (27) for the exchange strength n_{12} in terms of the functions $G(\mathbf{n})$ and $\hat{G}(\mathbf{n})$ and the associated HNC equations. Ignoring the elementary portions in these equations (HNC/0 approximation), we arrive at the result

$$v^*(\mathbf{n}) = \Delta(\mathbf{n}) - \frac{\varepsilon}{2}\left[N(\mathbf{n}) - \frac{G(\mathbf{n})}{1 - G^2(\mathbf{n})}\right] \quad . \qquad (36)$$

Relation (36) involves the nodal component of the distribution function $G(\mathbf{n})$, which has the decomposition (22) within the HNC classification scheme. The component $v_M^*(\mathbf{n})$ which contributes to the effective potential (35) only in the ordered phase, is given by

$$v_M^*(\mathbf{n}) = -\Delta(\mathbf{n}) + v_1(\mathbf{n}) + v_2(\mathbf{n}) + v_3(\mathbf{n}) + v_4(\mathbf{n}) \quad . \qquad (37)$$

To obtain explicit expressions for the potential (37), we must construct the variational derivative of the functional $e[G(\mathbf{n}), M, \lambda]$ representing the ground-state energy per spin with respect to the spatial distribution function $G(\mathbf{n})$, the order parameter M and the external parameter λ being held constant. The required algebraic manipulations are elementary but lengthy. The formal results for the components $v_i(\mathbf{n})$ ($i = 1, 2, 3, 4$) that contribute to the potential $v_M^*(\mathbf{n})$ of Eq. (37) read

$$v_1(\mathbf{n}) = \varepsilon \frac{G(\mathbf{n})}{1 + G(\mathbf{n})} \left[1 - \frac{2G(\mathbf{n})}{1 - G^2(\mathbf{n})}\right] + \varepsilon F(\mathbf{n}) G^2(\mathbf{n}) \frac{5 - G^2(\mathbf{n})}{[1 + G(\mathbf{n})]^2}$$

$$+ 2\varepsilon F(\mathbf{n}) \frac{G(\mathbf{n})}{[1 + G(\mathbf{n})]^2} \left[\frac{2G^2(\mathbf{n})}{1 - G(\mathbf{n})} - 3\right] G(\mathbf{n})(1 - M^2)^{-1} \quad , \tag{38}$$

$$v_2(\mathbf{n}) = \varepsilon W(\mathbf{n}) \left[1 - \hat{F}(\mathbf{n})^{-1} F(\mathbf{n})\right] + \varepsilon(1 - M^2) \left[\hat{F}(\mathbf{n})\hat{G}(\mathbf{n}) - \hat{N}(\mathbf{n})\right] \quad , \tag{39}$$

$$v_3(\mathbf{n}) = \varepsilon \sum_{\mathbf{m}} \left[G(\mathbf{m} - \mathbf{n}) + (1 - M^2)\hat{X}(\mathbf{m} - \mathbf{n})\right] V(\mathbf{m}) +$$

$$\varepsilon \sum_{\mathbf{m}} \left[G(\mathbf{m} - \mathbf{n}) + (1 - M^2)\hat{X}(\mathbf{m} - \mathbf{n})\right] \left[W(\mathbf{m}) - \hat{F}(\mathbf{m})\right]$$

$$\times \left[(1 - \delta_{\mathbf{m}0})\hat{F}(\mathbf{m})^{-1} - 1\right] \quad , \tag{40}$$

$$v_4(\mathbf{n}) = \varepsilon \sum_{\mathbf{m},\mathbf{l}} \left[W(\mathbf{m}) - \hat{F}(\mathbf{m})\right] M(\mathbf{m}, \mathbf{l}) W(\mathbf{n}, \mathbf{m}) \tag{41}$$

Expressions (38)-(41) involve the functions

$$F(\mathbf{n}) = \left[1 + \frac{1 - M}{1 + M} G(\mathbf{n})\right]^{-1} \left[1 + \frac{1 + M}{1 - M} G(\mathbf{n})\right]^{-1} \quad , \tag{42}$$

$$\hat{F}(\mathbf{n}) = \left[1 - M(1 + M)\hat{G}(\mathbf{n})\right]^{-1} \left[1 + M(1 - M)\hat{G}(\mathbf{n})\right]^{-1} \quad . \tag{43}$$

The potentials $V(\mathbf{n})$, $W(\mathbf{n})$ and $W(\mathbf{n}, \mathbf{m})$ are defined by

$$V(\mathbf{n}) = -\frac{1}{2}G(\mathbf{n}) + 1 - \hat{F}(\mathbf{n}) + \frac{1}{2}(1 - M^2)\left[\hat{N}(\mathbf{n}) + \hat{G}(\mathbf{n})\right]$$

$$-(1 - M^2)\hat{F}(\mathbf{n})\hat{G}(\mathbf{n}) \quad , \tag{44}$$

$$W(\mathbf{n}) = \frac{1}{2}(1 - M^2)\hat{N}(\mathbf{n}) - (1 - M^2)\hat{F}(\mathbf{n})\hat{G}(\mathbf{n}) + \sum_{\mathbf{m}} V(\mathbf{m})G(\mathbf{m} - \mathbf{n}), \tag{45}$$

and

$$W(\mathbf{n}, \mathbf{m}) = (1 - \delta_{\mathbf{n}0})\left[1 - F(\mathbf{n})\hat{F}(\mathbf{n})^{-1}\right]\delta_{\mathbf{n}\mathbf{m}}$$

$$+ \left[(1 - \delta_{\mathbf{n}0})\hat{F}(\mathbf{n})^{-1} - 1\right]\left[G(\mathbf{m} - \mathbf{n}) + (1 - M^2)\hat{X}(\mathbf{m} - \mathbf{n})\right]. \tag{46}$$

The kernel $M(\mathbf{n}, \mathbf{l})$ appearing in Eq. (41) is the solution of the linear equation

$$\sum_{\mathbf{l}} \left[\delta_{\mathbf{n}\mathbf{l}} + M(\mathbf{n}, \mathbf{l})\right]\left[\delta_{\mathbf{l}\mathbf{m}} - B(\mathbf{l}, \mathbf{m})\right] = \delta_{\mathbf{n}\mathbf{m}} \quad , \tag{47}$$

with the input quantity

$$B(\mathbf{l}, \mathbf{m}) = [(1 - \delta_{\mathbf{l0}})\hat{F}(\mathbf{l})^{-1} - 1]G(\mathbf{m} - \mathbf{l}) \quad . \tag{48}$$

In the strong-coupling limit ($\lambda \to 0$), the ground state is uncorrelated; hence, the quantities $G(\mathbf{n})$, $N(\mathbf{n})$, and $v_i(\mathbf{n})$ vanish, and the molecular-field approximation (16), (17) is exact. In the weak-coupling limit ($\lambda \to \infty$), the effective potential (35) specializes to $v(\mathbf{n}) = \Delta(\mathbf{n})$.

The quantity (34) may be interpreted as the derivative of a generalized spatial distribution function(Feenberg 1969)

$$\dot{G}(\mathbf{n}) = \frac{\partial}{\partial \alpha} G(\mathbf{n}, \alpha) \mid_{\alpha=0} \quad . \tag{49}$$

The function $G(\mathbf{n}, \alpha)$ can be generated from a pseudopotential $u(\mathbf{n}, \alpha) = u(\mathbf{n}) + \alpha v(\mathbf{n})$, in analogy with the generation of $G(\mathbf{n}, 0) \equiv G(\mathbf{n})$ from the pseudopotential $u(\mathbf{n}, 0) \equiv u(\mathbf{n})$. This identification allows us to derive a modified set of HNC equations for the function (34). Taking the derivative of the HNC equations for the generalized distribution function $G(\mathbf{n}, \alpha)$ with respect to the parameter α, we arrive at the decomposition

$$\dot{G}(\mathbf{n}) = \dot{X}(\mathbf{n}) + \dot{N}(\mathbf{n}) \tag{50}$$

and the HNC/0 equations (Ristig and Kim 1996)

$$\dot{X}(\mathbf{n}) = -\dot{N}(\mathbf{n}) + (1 - \delta_{\mathbf{n0}})\big[1 - G^2(\mathbf{n})\big]$$
$$\times \left[1 - M\frac{1 - G(\mathbf{n})}{1 + G(\mathbf{n})}\right]\left[1 + M\frac{1 - G(\mathbf{n})}{1 + G(\mathbf{n})}\right]\left[v(\mathbf{n}) + \frac{\dot{N}(\mathbf{n})}{1 - M^2}\right] \quad , \tag{51}$$

and

$$\dot{N}(\mathbf{n}) = \sum_{\mathbf{m}} \dot{X}(\mathbf{m} - \mathbf{n})G(\mathbf{m}) + \sum_{\mathbf{m}} X(\mathbf{m} - \mathbf{n})\big[\dot{X}(\mathbf{m}) + \dot{N}(\mathbf{m})\big] \tag{52}$$

for the non-nodal portion $\dot{X}(\mathbf{n})$ and the nodal component $\dot{N}(\mathbf{n})$.

The standard procedure(Campbell 1978), (Feenberg 1969) may be applied to solve the paired-magnon equation (33). We cast the Feynman equation (19) for the excitation energies of the magnons into the form of a renormalized Bogoliubov equation,

$$\omega^2(\mathbf{k}) = \varepsilon[\varepsilon + 2\nu(\mathbf{k})] \tag{53}$$

with

$$2\nu(\mathbf{k}) = \varepsilon S(\mathbf{k})^{-2}[1 - S(\mathbf{k})] + \varepsilon S(\mathbf{k})^{-1}[1 - S(\mathbf{k})] \quad , \tag{54}$$

and employ the Fourier transform of Eq. (33),

$$\dot{S}(\mathbf{k}) + \frac{\varepsilon}{2}(1 - M^2)[S(\mathbf{k}) - 1] = 0 \quad . \tag{55}$$

Insertion of relation (55) into the first term of expression (54) yields the particle-hole potential

$$\nu(\mathbf{k}) = \left\{(1 - M^2)^{-1}\dot{S}(\mathbf{k}) + \frac{\varepsilon}{2}S(\mathbf{k})[1 - S(\mathbf{k})]\right\}S^{-2}(\mathbf{k}) \quad . \tag{56}$$

To solve Eq. (53) numerically, we calculate the potential (56) and the excitation energy $\omega(\mathbf{k})$ via Eq. (53) and equate the result with the Feynman expression (19).

5   The Optimal Order Parameter

Variation of the energy functional (9) at constant external field and fixed spatial distribution function $G(\mathbf{n})$ yields an Euler-Lagrange equation for the optimal order parameter in the ordered phase,

$$\frac{\partial E}{\partial M^2} = 0 \quad . \tag{57}$$

The derivative (57) may be written as (Ristig and Kim 1996)

$$M^2 = 1 - \left(\frac{\lambda}{\Lambda}\right)^2 \quad , \tag{58}$$

where the Hartree field Λ depends on the magnetization itself and on the spatial distribution function $G(\mathbf{n})$. The ferromagnetic region $0 \leq \lambda \leq \lambda_c$ of the phase space is characterized by the inequality $\Lambda > \lambda$. In a mean-field approximation with $G(\mathbf{n}) \equiv 0$, the field Λ is constant at $\Lambda = 2D$. If spatial correlations are present, Λ is instead determined by the relation

$$\Lambda[G(\mathbf{n}), M, \lambda] = \frac{2D(1 + H_0)}{n_{12}(1 + H_1)} \tag{59}$$

from the energy components

$$H_0 = \frac{1}{2D} \sum_{\mathbf{n}} \Delta(\mathbf{n}) G(\mathbf{n}) \quad , \tag{60}$$

$$H_1 = -2(1 - M^2) \frac{\partial}{\partial M^2} \ln n_{12} \quad . \tag{61}$$

Employing the explicit formula (27) for the strength factor n_{12} in terms of the distribution function and the order parameter, we may express the potential H_1 as a functional of M and $G(\mathbf{n})$. The results of this derivation, in HNC/0 approximation may be summarized in the compact form

$$H_1 = (1 - M^2)(H^* + M^2 H_M) \quad . \tag{62}$$

In the strong-coupling regime ($\lambda \to 0$), quantities (60) and (61) vanish and the exchange strength approaches unity. Consequently, in this limit we recover the molecular-field result, $\Lambda = 2D$. Under increase of the coupling parameter λ, the field Λ decreases and becomes critical at $\Lambda = \lambda_c$, where $M = 0$.

Explicit expressions establishing the functional dependence of the potential H_1 may be derived by constructing the derivative of the spin-exchange strength (26) with respect to the magnetization M at fixed coupling parameter $0 \leq \lambda < \lambda_c$ and at fixed distribution function $G(\mathbf{n})$. The components H^* and H_M contributing to the potential (62) are given by the expressions

$$H^* = - \sum_{\mathbf{n}} [F(\mathbf{n}) G(\mathbf{n}) (1 - M^2)^{-1} + \delta_{\mathbf{n}0} \hat{X}(\mathbf{n})]$$

$$-\sum_{\mathbf{n}}\hat{X}(\mathbf{n})\big[1+\hat{X}(\mathbf{n})\big]-\sum_{\mathbf{n}}[G(\mathbf{n})+\hat{X}(\mathbf{n})]\hat{N}(\mathbf{n}) \quad , \tag{63}$$

$$
\begin{aligned}
H_M = &-\sum_{\mathbf{n}}F(\mathbf{n})G^2(\mathbf{n})(1-M^2)^{-2}+\sum_{\mathbf{n}}[1-\hat{F}(\mathbf{n})]M^{-2}\hat{G}(\mathbf{n})\big[1+\hat{G}(\mathbf{n})\big]\\
&-\sum_{\mathbf{n}}\hat{N}(\mathbf{n})G(\mathbf{n})(1-M^2)^{-1}+\sum_{\mathbf{n}}\hat{H}(\mathbf{n})P(\mathbf{n})\\
&+\sum_{\mathbf{n}}\hat{G}(\mathbf{n})\big[3\hat{F}(\mathbf{n})\hat{G}(\mathbf{n})+2\hat{F}(\mathbf{n})G(\mathbf{n})(1-M^2)^{-1}-2\hat{N}(\mathbf{n})\big]\\
&+\sum_{\mathbf{n},\mathbf{m}}\big[(1-\delta_{\mathbf{n0}})\hat{F}(\mathbf{n})+(1-M^2)\hat{H}(\mathbf{n})\big](1-M^2)^{-1}M(\mathbf{n},\mathbf{m})P(\mathbf{m})
\end{aligned}\tag{64}
$$

The functions $P(\mathbf{n})$ and $\hat{H}(\mathbf{n})$ are defined by

$$
\begin{aligned}
P(\mathbf{n}) = &(1-M^2)\hat{G}^2(\mathbf{n})+2(1-M^2)^{-1}G(\mathbf{n})\big[\hat{F}(\mathbf{n})^{-1}F(\mathbf{n})-1\big]\\
&-(1-\delta_{\mathbf{n0}})\hat{F}(\mathbf{n})^{-1}\hat{X}(\mathbf{n})\\
&+(1-M^2)\hat{G}(\mathbf{n})M^{-2}\big[1-\hat{F}(\mathbf{n})^{-1}\big]\\
&+(1-\delta_{\mathbf{n0}})\hat{F}(\mathbf{n})^{-1}M^{-2}\big[\hat{X}(\mathbf{n})+F(\mathbf{n})G(\mathbf{n})+F(\mathbf{n})G^2(\mathbf{n})\big],
\end{aligned}\tag{65}
$$

$$
\begin{aligned}
\hat{H}(\mathbf{n}) = &(1-\delta_{\mathbf{n0}})\big[\hat{F}(\mathbf{n})\hat{G}(\mathbf{n})-\tfrac{1}{2}\hat{N}(\mathbf{n})\big]\\
&+\sum_{\mathbf{m}}(1-\delta_{\mathbf{m0}})\big[\hat{F}(\mathbf{m})\hat{G}(\mathbf{m})-\tfrac{1}{2}\hat{N}(\mathbf{m})-\tfrac{1}{2}\hat{G}(\mathbf{m})\big]G(\mathbf{m}-\mathbf{n})\\
&+\sum_{\mathbf{m}}\big[(1-\delta_{\mathbf{m0}})\hat{F}(\mathbf{m})-1+\tfrac{1}{2}G(\mathbf{m})\big](1-M^2)^{-1}G(\mathbf{m}-\mathbf{n}) \quad .
\end{aligned}\tag{66}
$$

6 Numerical Results and Discussion

Employing the variational CBF formalism in conjunction with the HNC/0 approximation, we have calculated the optimal spatial distribution function $G(\mathbf{n})$, the optimal magnetization M, and various physical quantities associated with the transverse Ising model of spins on a square lattice, as functions of the coupling parameter λ. The calculations are based on ansatz (6)-(8) and the Euler-Lagrange equations (33) and (58) derived for an infinitely extended lattice. Equations (15) and (53) are used for practical evaluation of the optimal distribution function or static structure function (21) within a standard iteration scheme (Campbell 1978). Numerical solution of the relevant equations is carried out for an $(L \times L)$ lattice with $L = 8$ and $L = 16$, assuming periodic boundary conditions. The calculations are performed at coupling parameters λ within the entire range $(0 \leq \lambda \leq \infty)$. The exact results for the model are reproduced both in the strong-coupling limit $(\lambda \to 0)$ and in the weak-coupling region $(\lambda \to \infty)$. Numerical results are reported for optimal gross quantities including the ground-state energy and its potential component, the magnetization (3), the spin-exchange strength (14). In this section we analyze the properties of the optimal distribution function, the structure function, and the correlation length as functions of the coupling parameter. We further present some of our results for the optimal excitation energies of the magnons and give special attention to the properties of the optimized quantities in the phase transition region.

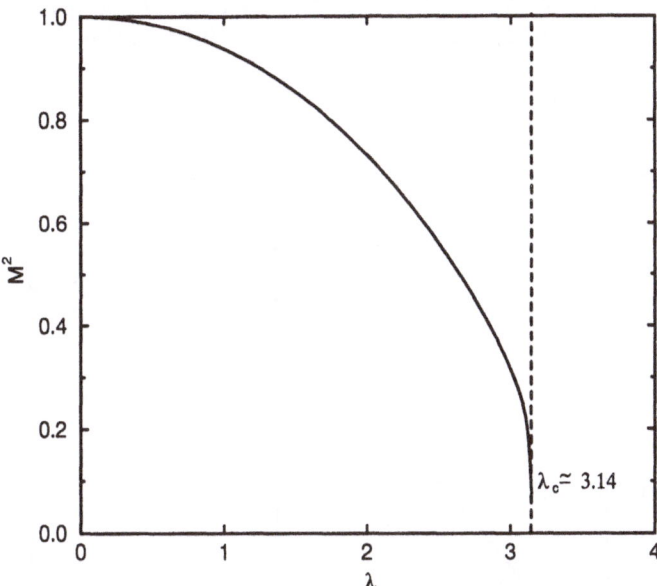

Fig. 1. CBF results for the square of the optimal order parameter (magnetization M) in the ferromagnetic phase, in HNC/0 approximation. The long-range order vanishes at the critical coupling parameter $\lambda_c \simeq 3.14$. The function $M^2(\lambda)$ behaves non-analytically at the critical point due to the divergence of the correlation length that characterizes the spatial distribution function (cf. Eq. (72)).

Fig. 1 displays the numerical results from the solutions of Eq. (58) that determine the square of the optimal order parameter as a function of the external field λ. The magnetization M decreases from unity with increase of the coupling parameter λ and vanishes at the critical value $\lambda = \lambda_c \simeq 3.14$. In the disordered phase, M is of course identically zero. We find good quantitative agreement with the numerical predictions by Pfeuty and Elliott (Pfeuty and Elliott 1971), which are believed to be the best currently available from a first-principles treatment(Bishop, Kendall, Wong, and Xian 1993). We note that the calculated M^2

vanishes with an infinite slope, in contrast to the behavior of the molecular-field result, which shows a finite slope as $M^2 \to 0$. The singularity is caused by the long-range behavior of the optimal spatial distribution function $G(\mathbf{n})$ as the critical field $\Lambda = \lambda_c$ is approached in the ordered region (see further remarks below). This long-range effect is also reflected in the properties of the optimal Hartree field Λ, as a function of the parameter λ in the ordered phase ($\lambda \leq \lambda_c$). The mean-field approximation $\Lambda \simeq 2D$ (with $D = 2$ in the present study) correctly describes the strong-coupling limit ($\lambda \to 0$), but, with increasing λ, deviates rapidly from the prediction that takes account of the spatial correlations present at nonzero coupling. Our numerical results approach the critical point ($\lambda \to \lambda_c = \Lambda \simeq 3.14$) continuously but with infinite slope.

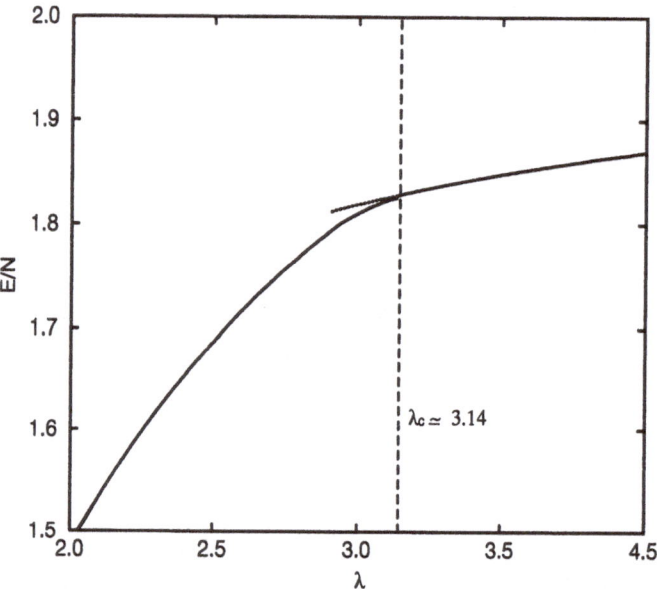

Fig. 2. CBF results for the minimized energy per spin, in HNC/0 approximation, as a function of the coupling parameter λ. As $\lambda \to 0$, the mean-field behavior given by Eq. (17), exact in this limit, is correctly reproduced. In the asymptotic region ($\lambda \to \infty$), the CBF result agrees with the result from second-order perturbation theory, $E/N \simeq 2 - (2\lambda)^{-1}$. The energy depends smoothly on the external field λ in the critical region where the continuous phase transition occurs. The metastable energy branch ($M \equiv 0, \lambda < \lambda_c$) is indicated by the dotted line.

Fig. 2 presents our numerical results on the minimal ground-state energy per spin (or lattice site) in the ordered and disordered phases. The energy increases monotonically from the molecular-field result (16) valid for $\lambda \ll \lambda_c$, and, at the critical point λ_c, smoothly joins the numerical results obtained for the disordered phase. Imposing the condition $M \equiv 0$ at parameter values $\lambda \leq \lambda_c$, the Euler-Lagrange equations admit the existence of metastable solutions which have energies larger than those of the stable ferromagnetic branch but do not break the symmetry prevailing in the disordered phase (indicated by the dotted curve in Fig. 3). This metastable branch eventually becomes unstable at sufficiently small λ values (somewhere below $\lambda \simeq 2.8$), where the paramagnons soften (see subsequent discussion).

The potential energy (10) and the magnetic energy (11) are given by

$$V = ND(1 - M^2)(1 + H_0) \quad , \tag{67}$$

$$V_M = N\lambda(1 - A) \quad . \tag{68}$$

The energy (68) represents essentially the transverse magnetization A. We note further that this quantity is closely related to the total derivative of the energy (9) with respect to the coupling parameter in the optimal state $|\Psi\rangle$, since

$$\frac{1}{N}\frac{dE(\lambda)}{d\lambda} = \frac{\partial}{\partial\lambda}e[G(\mathbf{n}), M, \lambda] = 1 - A \quad . \tag{69}$$

The quantity (67) may be identified with the analog of the entropy of statistical mechanics(Hamer and Kogut 1979),

$$S = \frac{d}{dx}\big[xE(x)\big] = E - \lambda\frac{dE}{d\lambda} = V \quad , \tag{70}$$

with $x = \lambda^{-1}$, assuming the state $|\Psi\rangle$ has been optimized.

The results for the optimized functions (67) and (68) depend continuously on the coupling parameter λ but develop a divergent slope as the critical point λ_c is approached from the ordered region. The potential energy (Fig. 3) and the magnetic energy decrease rather slowly with increasing coupling parameter λ in the paramagnetic phase. The transverse magnetization A is correctly described by the mean-field approximation $A \simeq \lambda/2D$ as $\lambda \to 0$ and becomes unity in the weak-coupling limit ($\lambda \to \infty$). The results for the energy portion (67) in the metastable region $\lambda < \lambda_c$ characterized by $M \equiv 0$ are indicated in Figs. 3 by the dotted line.

The numerical results for the optimized spin-exchange strength (14) are plotted in Fig. 4. This function approaches unity in the limits $\lambda \to 0$ and $\lambda \to \infty$, since the spin system is uncorrelated in these regions of the phase space. In the intermediate region, the function $n_{12}(\lambda)$ is smaller than unity. The stable phases are continuously connected, but the strength function exhibits a non-analytic cusp at the transition point λ_c. The metastable branch of the disordered phase ($\lambda < \lambda_c, M \equiv 0$) is characterized by a suppressed spin exchange that is too weak to break the symmetry of the paramagnetic system (dotted curve).

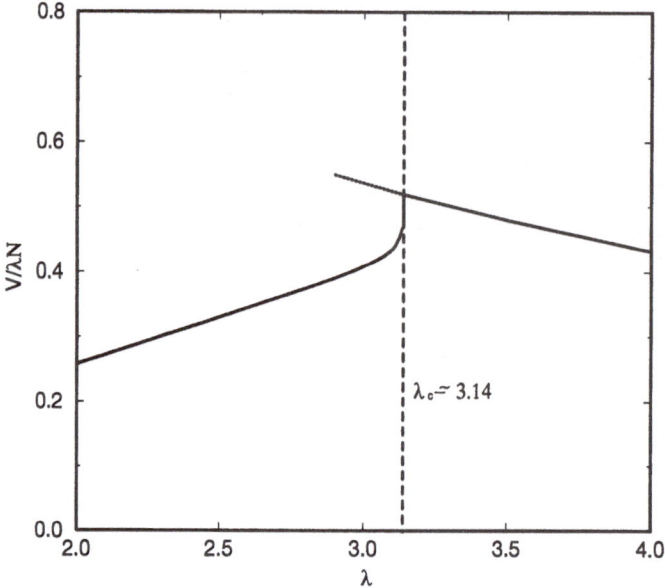

Fig. 3. The optimal potential component (67) per spin divided by λ, as a function of the coupling constant, in HNC/0 approximation. The second-order transition occurs at the critical point $\lambda_c \simeq 3.14$. Since the ferromagnon excitations become soft at the phase boundary, the curve representing the energy component develops an infinite slope at λ_c. The results for the metastable branch are given by the dotted curve.

The singular behavior of the optimized physical quantities at the ferromagnetic boundary, predicted by the variational CBF analysis, is most clearly expressed by the properties of the total derivative of the transverse magnetization $A(\lambda)$ with respect to the parameter λ. This derivative is essentially the analog of the specific heat familiar from statistical physics. Following Hamer and Kogut (Hamer and Kogut 1979), we define

$$c_\lambda = -\frac{1}{N}\frac{dS}{dx} = -\frac{1}{N}\frac{d^2}{dx^2}\big[xE(x)\big] = \lambda^3 \frac{dA(\lambda)}{d\lambda} \quad . \tag{71}$$

In the strong-coupling regime ($\lambda \ll \lambda_c$), the results are correctly represented by the molecular-field approximation, $c_\lambda \simeq (2D)^{-1}\lambda^3$. In the asymptotic region ($\lambda \to \infty$), the quantity (71) approaches unity. On the ferromagnetic side of the transition, c_λ diverges rapidly as the critical point λ_c is approached. By contrast, the dependence of c_λ on the coupling parameter is rather weak in the disordered state and extends smoothly into the metastable region.

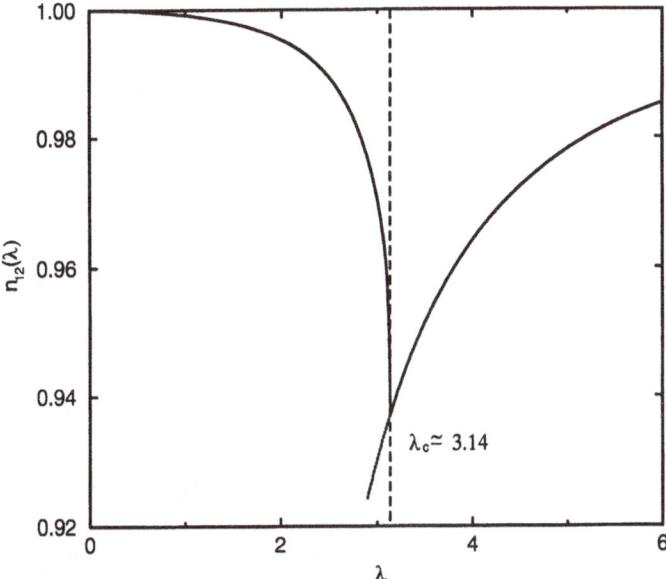

Fig. 4. Numerical results, in HNC/0 approximation, for the optimal spin-exchange strength (14), as a function of the coupling parameter λ in the ordered and disordered phases. The dotted line represents the results for the metastable region of the disordered phase.

Any non-mean-field aspects of the gross properties of the transverse Ising model displayed in Figs. 1-4 originate, of course, from the spatial correlations characterized by the modified distribution function $G(\mathbf{n})$ of Eq. (15) or the corresponding structure function (21). The numerical results for the function $G(\mathbf{n})$ obtained within the current CBF analysis are well represented by the formula

$$G(\mathbf{n}) \simeq \frac{G_0}{|\mathbf{n}|} e^{-\kappa |\mathbf{n}|} \quad , \tag{72}$$

where the inverse correlation length κ depends on the strength of the transverse magnetic field. Such a behavior has been already observed in a study of the two-dimensional $U(1)$ lattice gauge model(Dabringhaus and Ristig 1991). The optimal results for the correlation length $\kappa^{-1}(\lambda)$ increase monotonically as the transition point is approached. While this quantity diverges in the ferromagnetic phase as $\lambda \to \lambda_c$, the spatial distribution function $G(\mathbf{n})$ remains of short range if the critical point λ_c is approached from the stable disordered region. By locking the system into a state of high symmetry ($M \equiv 0$), we may therefore extend the disordered state into a metastable region ($\lambda < \lambda_c$).

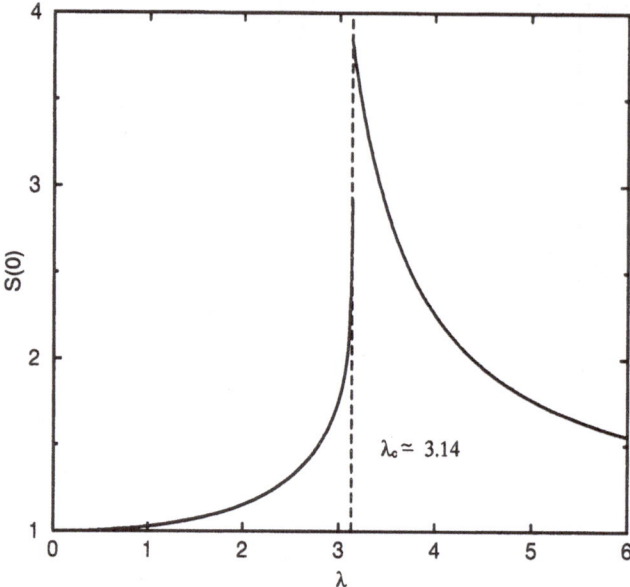

Fig. 5. CBF results for the optimal static structure function $S(\mathbf{k})$ at zero wave vector. This quantity diverges as $\lambda \to \lambda_c$ from below.

As evidenced by Fig. 5, the static structure function (21) at vanishing wave vector $\mathbf{k}$ reflects this behavior of the optimal spatial distribution function. The spins are uncorrelated in the limits $\lambda \to 0$ and $\lambda \to \infty$, and thus $S(\mathbf{k})$ is unity at $\mathbf{k} = 0$. The quantity $S(0)$ increases with increasing strength λ of the external field in the ferromagnetic phase and ultimately develops a divergence as the critical point λ_c is approached. The expected smooth dependence of the structure factor $S(0)$ on the coupling parameter λ is found in the stable and metastable regions of the paramagnetic phase.

Fig. 6 presents our numerical results for the dispersion relation of the magnon excitations at zero momentum. The data on the optimized energy gap $\omega(0)$ recover the exact results $\omega(0) = 4D$ and $\omega(0) = 2\lambda$ associated with the true excited states of the Hamiltonian (1) in the limits $\lambda \to 0$ and $\lambda \to \infty$, respectively. The excitation energy is drastically reduced in the transition region due to the spatial correlations in the system. Since the correlation length $\kappa^{-1}(\lambda)$ diverges in the

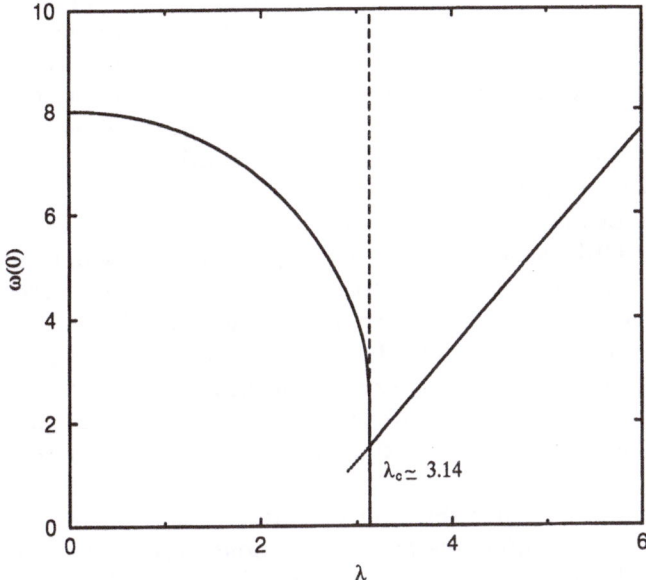

Fig. 6. The magnon energy gap as a function of the coupling parameter λ. Plotted are the numerical results for the optimized excitation energy, in Feynman approximation (19). The gap associated with the ferromagnons vanishes at the critical point $\lambda_c \simeq 3.14$. The paramagnons in the metastable region $\lambda \leq \lambda_c$ have small but nonzero energies (dotted line).

ordered phase if $\lambda \to \lambda_c$, the gap $\omega(0)$ vanishes in this case, and consequently the ferromagnons become soft at the transition point. In contrast, paramagnons may (theoretically) exist in a metastable disordered phase, but are eventually unstable at sufficiently small values of the coupling parameter ($\lambda < 2.8$).

In summary, CBF theory at its variational level has been adapted to treat the ground and excited states of the vacuum sector of the abelian $Z(2)$ lattice gauge model defined on an infinitely extended square lattice. Duality has been used to formulate the treatment within the physical picture of an Ising model of N spins in a transverse magnetic field, with $N \to \infty$ and fixed lattice constant. Two Euler-Lagrange equations have been derived to determine the optimal distribution function and the optimal order parameter which characterize the spatial spin correlations and the ordered structure of the system. HNC equations (in HNC/0 approximation) have been employed to relate these quantities to a correlated many-body wave function of Hartree-Jastrow type. A detailed numerical application of the theory has been carried out within this optimized variational version of CBF theory. The actual numerical solutions of the Euler-Lagrange equations have been obtained for a 8×8 and a 16×16 square lattice with periodic boundary conditions. Finite-size effects are quite small and may be ignored as long as we do not probe the fine structure of the model in the ferromagnetic region $|\lambda - \lambda_c| < 0.05$ very close to the critical point. We stress that our primary aim is a reliable quantitative study outside of the immediate vicinity of the transition. In that narrow region, scaling theories of critical exponents are more appropriate than the approach pursued here. Accordingly, we have not attempted to extract detailed information on critical exponents within the present formulation of CBF theory. We have reported results on the optimized structure function, and on various gross properties such as the minimum ground-state energy, its potential and magnetic components, the magnetization, and the analog of the specific heat. We have also furnished numerical information on the dispersion relation of the elementary excitations. In particular, we have shown that the CBF analysis yields an accurate prediction for the critical point and enables us to bridge the gap between perturbative methods and scaling theories of phase transitions.

In closing, we point out that the CBF formalism may be adapted to provide microscopic *ab initio* treatments of various other correlated many-body systems or models of interest such as the charged sector of abelian lattice gauge models(Horn, Weinstein, and Yankielowicz 1979), gauge models with matter fields(Irving and Thomas 1982), antiferromagnetic or frustrated systems, more complex spin-models(Bishop, Parkinson, and Xian 1992), and pseudospin systems describing ferroelectric crystals and structural phase transitions(Binder 1991), (Lüthi and Rehwald 1981).

Acknowledgments

This work has been supported in part by the Deutsche Forschungsgemeinschaft (Graduiertenkolleg on "Classification of phase transitions in crystalline materials") and under grant Nos. Ri 267/26/27, by the EC Human Capital and Mobility Program under Contract No. ERBCHRXCT 940456, and by the U. S. National Science Foundation under Grant No. PHY-9602127.

References

Allton C. R., Yung C. M., and Hamer C. J., Phys. Rev. **D 39**, 3772 (1989).

Binder K., in *Materials Science and Technology*, Vol. 5, *Phase Transformations in Materials*, edited by Haasen P. (VCH Verlag, Weinheim, 1991).

Bishop R. F., Kendall A. S. , Wong L. Y., and Xian Y., Phys. Rev. **D 48**, 887 (1993).

Bishop R. F., Kendall A. S., Wong L. Y., and Xian Y., in *Condensed Matter Theories*, edited by Blum L. and Malik F.B. (Plenum, New York, 1993) Vol. 8.

Bishop R. F., in *Recent Progress in Many-Body Theories*, edited by Schachinger E., Mitter H., and Sormann H. (Plenum, New York, 1995) Vol. 4.

Bishop R. F., Davidson N. J., and Xian Y., in *Recent Progress in Many-Body Theories*, edited by Schachinger E., Mitter H. , and Sormann H. (Plenum, New York, 1995) Vol. 4.

Bishop R. F., Parkinson J. B., and Xian Y., Phys. Rev. **B 43**, 13782 (1991); **B 44**, 9425 (1991); **B 46**, 880 (1992).

Campbell C. E., in *Progress in Liquid Physics*, edited by Croxton C. A. (Wiley, New York, 1978).

Clark J. W. and Feenberg E., Phys. Rev. **113**, 388 (1959).

Clark J. W., in *Progress in Particle and Nuclear Physics*, edited by Wilkinson D. (Pergamon, Oxford, 1979), Vol. 2.

Clements B. E., Forbert H., Krotscheck E., Lauter H. J., Saarela M., and Tymczak C. J., Phys. Rev. **B 50**, 6958 (1994).

Creutz M. , Jacobs L., and Rebbi C., Phys. Rep. **95**, 201 (1983).

Dabringhaus A., Ristig M. L., and Clark J. W., Phys. Rev. **D 43**, 1978 (1991).

Dabringhaus A. and Ristig M. L., in *Condensed Matter Theories*, edited by Fantoni S. and Rosati S. (Plenum, New York, 1991) Vol. 6.

Dagotto E. and Moreo A., Phys. Rev. **D 29**, 300 (1984).

Dagotto E., Masperi L., Moreo A., Della Selva A., and Fiore R., Phys. Rev. **D 32**, 1491 (1985).

Di Bartolo C., Gambini R., and Trias A., Phys. Rev. **D 39**, 3136 (1989).

Feenberg E., *Theory of Quantum Fluids* (Academic, New York, 1969).

Feynman R. P., Phys. Rev. **94**, 262 (1954).

Fradkin E. and Raby S., Phys. Rev. **D 20**, 2566 (1979).

Gernoth K. A., this volume.

Gernoth K. A. , Clark J. W., Senger G., and Ristig M. L., Phys. Rev. **B 49**, 15 836 (1994).

Gernoth K. A., Clark J. W. and Ristig M. L., Z. Physik **B 98**, 337, 1995.

Hamer C. J. and Irving A. C., J. Phys. **A 17**, 1649 (1984).

Hamer C. J. and Kogut J. B., Phys. Rev. **B 20**, 3859 (1979).

Horn D., Weinstein M., and Yankielowicz S., Phys. Rev. **D 19**, 3715 (1979).

Irving A. C. and Thomas A., Nucl. Phys. **B 200**, 424 (1982).

Kramers H. A. and Wannier G. H., Phys. Rev. **60**, 252 (1941).

Lüthi B. and Rehwald W., in *Topics in Current Physics*, Vol. **23**, *Structural Phase Transitions I* , edited by Müller K. A. and Thomas H. (Springer-Verlag, Berlin, 1981).

Penson K. A., Jullien R., and Pfeuty P., Phys. Rev. **B 19**, 4653 (1979).

Pfeuty P. and Elliott R. J., J. Phys. **C 4**, 2370 (1971).

Ristig M. L. and Kim J. W., Phys. Rev. **B 53**, 6665 (1996).

Ristig M. L., Fantoni S., and Kürten K. E., Z. Physik **B 51**, 1 (1983).

Ristig M. L., Kim J. W., and Mehlmann R., in *Condensed Matter Theories*, edited by Ludeña E. et. al. (Nova Science Publishers, Commack, NY, 1996) Vol. 11.

Ristig M. L., in *From Nuclei to Particles, Proceedings of the International School of Physics Enrico Fermi, Course LXXIX, Varenna 1981*, edited by Molinari A. (North-Holland, Amsterdam, 1982).

Roomany H.H. and Wyld H. W., Phys. Rev. **D 21**, 3341 (1980).

Rushbrooke G. S., Baker G. A., and Wood P. J., in *Phase Transitions and Critical Phenomena*, edited by Domb C. and Green M. S. (Academic, London, 1974) Vol. 3.

Savit R., Rev. Mod. Phys. **52**, 453 (1980).

Stinchcombe R. B., J. Phys. **C 6**, 2459 (1973); 2507 (1973).

Crystallographic Point and Space Group Symmetries in Correlated Many-Body Wave Functions

K. A. Gernoth

Institute for Theoretical Physics, University of Cologne, D-50937 Cologne, Germany.
Department of Physical Sciences, Theoretical Physics, University of Oulu, FIN-90570 Oulu, Finland.
International School for Advanced Studies (SISSA/ISAS), I-34014 Trieste, Italy

1 Introduction

Spatially inhomogeneous phases of strongly correlated quantum many-body systems persist to provide a demanding challenge to the various many-body techniques that today are at the disposal of theoretical physicists working in this field. In particular, during the last fourteen years or so, significant progress has been made in generalizing and applying microscopic correlated many-body wave function theories to non-uniform Bose (Saarela, Pietiläinen, and Kohn 1983), (Krotscheck, Qian, and Kohn 1985a), (Krotscheck 1985), (Szybisz and Ristig 1989), (Gernoth, Clark, Senger, and Ristig 1994a), (Clements, Krotscheck, and Tymczak 1996) and Fermi (Krotscheck 1985c) systems. Beyond, correlated many-body density matrix theories that have proven successful for treating isotropic Bose fluids (Campbell, Kürten, Ristig, and Senger 1984), (Senger, Ristig, Kürten, and Campbell 1986) have been extended and applied to symmetry-broken, inhomogeneous Bose systems at elevated temperature (Gernoth, Clark, and Ristig 1995), (Gernoth, Clark 1994b), (Campbell, Clements, Krotscheck, and Saarela). For the helium quantum systems in particular these advances continued to be paralleled by similar developments in density functional theory (Stringari and Treiner), (Guirao, et. al. 1992), otherwise extensively applied especially to correlated electrons in atoms, molecules, solids, and alloys (March 1992), (Alonso and March 1989), and by progress achieved with diffusion (Chin and Krotscheck 1992), path integral (Wagner and Ceperley 1994), and variational Monte Carlo methods (Pederiva, Ferrante, Fantoni and Reatto 1995), (Pederiva, et. al. 1996). A list of physical systems that are bound to remain of prime experimental and theoretical interest and that may be tackled with such many-body theoretical tools would include the correlated electron liquid in crystalline solids (Krotscheck 1985c), (Chin and Krotscheck 1992), (Fulde 1991), surfaces and interfaces of quantum fluids and solids (Gernoth, Clark, and Ristig 1995), (Pederiva, et. al. 1994), adsorbed helium films with their wealth of diverse phases, transition phenomena, and excitations (Clements, Krotscheck, and Tymczak 1996), (Campbell, Clements, Krotscheck, and Saarela), magnetic systems, which theoretically are often described with spin lattice models (Bishop, Parkinson, and Xian 1993), (Ristig and Kim 1996), and the various crystalline phases of magnetic and non-magnetic quantum solids (Pederiva, Ferrante, Fantoni and Reatto 1995), (Pederiva, et. al. 1996).

Focusing on the one-body and pair correlation level, the present research is devoted to a presentation of how the paraphernalia furnished by the theory of point and space groups and their irreducible representations may be harnessed for correlated many-body wave function theories in the case of single-component, condensed systems exhibiting strictly crystalline symmetries. Proceeding from the foundation laid here, these group-theoretical methods may be suitably extended to cover also multi-component systems as well as higher-body effects. In this work the many-Fermion system will be discussed only in the limits of a polarized or unpolarized state. To treat the general many-Fermion state at the same level of rigor that is achieved for Bosons within non-magnetic space group theory would necessitate taking resort to the correspondingly more demanding theory of magnetic point and space groups and their irreducible corepresentations (Bradley and Cracknell 1972). As compared to the 230 non-magnetic crystallographic space groups in three dimensions and 17 in two dimensions, the sheer number of 1651 magnetic crystal space groups in three dimensions and 92 plane ones is impressive. The group-theoretical framework developed here in the context of purely crystalline point and space group symmetries promises to be of value also for a systematic extension of microscopic many-body theories to magnetic systems of more complex symmetry than can be handled at present. It is useful that such attempts are preceded by an exploration of opportunities afforded to many-body physics by non-magnetic space group theory.

The outline of the paper is as follows. In Section 2 the generic concept of a crystallographic space group is introduced, both in configuration space as well as in function (Hilbert) spaces. The main features of symmorphic and non-symmorphic space groups are briefly surveyed. The reader is referred to Ref.(Burns and Glazer 1990) for a comprehensive treatment of space groups and to Ref.(Cornwell 1984) for the group-algebraic techniques employed in this work. Ground and excited states of correlated quantum liquids and crystals, such as the helium crystals or the correlated electron liquid in solids, may suitably be described by many-body wave functions that appropriately incorporate the space group symmetries of the particular material under investigation. Such ground state many-body wave functions and the most basic consequences resulting from symmetry principles are discussed in Section 3. In Section 4 the general group-theoretical methods are outlined by means of which space group symmetries may correctly be taken into account in constructing suitable forms for the one-body and two-body correlation factors in the ground state wave function of a many-Boson or many-Fermion quantum system. Group-theoretical techniques facilitate also the derivation of the general shape of these one-body and two-body contributions in the form of expansions in terms of complete sets of symmetrized basis functions of variables in $|R^3$ and $|R^6$, respectively. Specifically, in Section 5, space group $P6_3/mmc$ of the hexagonal close-packed structure, in which, e.g., helium, beryllium, or magnesium crystallize, is discussed in detail with particular attention being devoted to elucidating the consequences of the non-symmorphic nature of the 6_3 screw axis and of the three vertical c-glide mirror planes.

The correlated ground state wave functions arrived at in the manner sketched above are suited for variational quantum Monte Carlo calculations and also for discussing order parameter symmetries of phase transitions from a quantum liquid to a quantum solid and between crystalline phases of quantum solids. The group-theoretical framework presented here is inspired by the use of similar expansion schemes for the probability densities in the treatment of phase transitions in liquid crystals and by tools developed for analyzing phase transitions from an isotropic liquid phase to a crystalline phase and might prove useful for studying also such phenomena (Tolédano and Tolédano 1987). For variational Monte Carlo simulations the coefficients of expansions of single-particle and pair correlation terms in complete sets of symmetry-adapted basis functions must be interpreted as variational parameters that may be determined by minimizing the variance of the energy expectation value. Monte Carlo techniques for optimizing a variational many-body wave function containing a large number of variational parameters are available and have been applied successfully in conjunction with expansions in complete sets of purely radial basis functions to quantum systems with isotropic correlations (Vitiello and Schmidt 1992), (Moroni, Fantoni, and Senatore 1995). Group-theoretical methods as described here may be employed to identify the space group resulting from the spontaneous symmetry-breaking found in variational Monte Carlo calculations with shadow wave functions (Pederiva, Ferrante, Fantoni and Reatto 1995). The group-theoretical formalism developed in this work may be expected to become of value also for the variational Euler-Lagrange approach to crystalline phases of correlated quantum systems (Krotscheck 1990) and to collective excitations and their symmetry-breaking nature relative to the parent ground state symmetry (Saarela and Suominen 1989), (Saarela 1995).

2 Crystallographic Space Groups

A crystallographic space group $\mathcal{G}$ consists of configuration space transformations $\{R, \mathcal{T}_R + \mathbf{t}\}$ in three-dimensional space $|\mathrm{R}^3$ defined by their action

$$\{R, \mathcal{T}_R + \mathbf{t}\}\mathbf{r} = R\mathbf{r} + \mathcal{T}_R + \mathbf{t} \tag{1}$$

on the three-dimensional coordinate vectors $\mathbf{r} \in |\mathrm{R}^3$. The symbol $\mathbf{t}$ denotes a vector

$$\mathbf{t} = n_1 \mathbf{a}_1 + n_2 \mathbf{a}_2 + n_3 \mathbf{a}_3 \tag{2}$$

of the direct Bravais lattice associated with $\mathcal{G}$. The numbers n_1, n_2, and n_3 may assume any positive, negative, or vanishing integral values. The vectors $\mathbf{a}_1$, $\mathbf{a}_2$, and $\mathbf{a}_3$ are basic lattice vectors spanning a primitive unit cell of the direct Bravais lattice. A (proper or improper) rotation of the crystallographic point group $\mathcal{G}_0$ of space group G is denoted by R. With $\mathbf{r} = (x_1, x_2, x_3) \in |\mathbb{R}^3$, the i-th component $(R\mathbf{r})_i$, $i = 1, 2, 3$, of the rotated vector $R\mathbf{r} = \big((R\mathbf{r})_1, (R\mathbf{r})_2, (R\mathbf{r})_3\big)$ is given by

$$(R\mathbf{r})_i = \sum_{j=1}^{3} R_{ij}\, x_j\,, \quad i = 1, 2, 3\,. \tag{3}$$

The point group G_0 of a space group G comprises all distinct rotational parts R occurring in the space group operations $\{R, \mathcal{T}_R + \mathbf{t}\} \in G$.

As indicated by the subscript R, the non-primitive translations $\mathcal{T}_R$ in (1) depend on the rotational part $R \in G_0$ of the point group G_0 of space group G and, of course, on the given space group G. The non-primitive translations $\mathcal{T}_R$ are unique up to Bravais lattice vectors and, for the sake of uniqueness, may always be chosen in a such a way that

$$\mathcal{T}_R = q_1 \mathbf{a}_1 + q_2 \mathbf{a}_2 + q_3 \mathbf{a}_3 \quad \text{with} \quad 0 \leq q_1, q_2, q_3 < 1\,. \tag{4}$$

The numbers q_i, $i = 1, 2, 3$, may assume only certain rational values, subject to the given constraints, these values depending on the particular space group G and on the particular rotational part $R \in G_0$ of space group G that one is dealing with.

In the Seitz notation for space group elements as introduced in definition (1) group multiplication is defined by the relation

$$\{R_1, \mathcal{T}_{R_1} + \mathbf{t}_1\}\{R_2, \mathcal{T}_{R_2} + \mathbf{t}_2\} = \{R_1 R_2, R_1 \mathcal{T}_{R_2} + R_1 \mathbf{t}_2 + \mathcal{T}_{R_1} + \mathbf{t}_1\}\,, \tag{5}$$

yielding also the useful expression

$$\{R, \mathcal{T}_R + \mathbf{t}\}^{-1} = \{R^{-1},\, -R^{-1}\mathcal{T}_R - R^{-1}\mathbf{t}\} \tag{6}$$

for the inverse $\{R, \mathcal{T}_R + \mathbf{t}\}^{-1}$ of a space group transformation $\{R, \mathcal{T}_R + \mathbf{t}\} \in G$. The group

$$\mathbf{T} = \Big\{\{1, \mathbf{t}\}\,\Big|\,\mathbf{t} \text{ a Bravais lattice vector}\Big\} \subseteq G \tag{7}$$

of pure primitive translations $\{1, \mathbf{t}\}$ by Bravais lattice vectors $\mathbf{t}$, the '1' denoting the identity element of point group G_0, is an invariant subgroup of space group G.

In the case of one of the **73 symmorphic** space groups the origin of the co-ordinate frame of reference, in which the space group symmetry transformations $\{R, \mathcal{T}_R + \mathbf{t}\}$ are defined, may be chosen such that all $\mathcal{T}_R$ vanish simultaneously, i.e.

$$\mathcal{T}_R = 0\,, \quad \forall R \in G_0\,. \tag{8}$$

A symmorphic space group G is the semi-direct product of the group $\mathbf{T}$ of primitive lattice translations with the point group G_0 of space group G,

$$G = \mathbf{T} \otimes_s G_0\,. \tag{9}$$

With a choice of origin as mentioned above, the point group $\mathcal{G}_0$ is contained in space group $\mathcal{G}$ as a proper subgroup,

$$\mathcal{G}_0 \subset \mathcal{G}. \tag{10}$$

In the case of one of the remaining **157 non-symmorphic** space groups, containing true screw axes and/or glide mirror planes, no origin of the coordinate frame of reference can be found so that all non-primitive translations $\mathcal{T}_R$ vanish. Certain rotations R of the point group $\mathcal{G}_0$ of space group $\mathcal{G}$ make their appearance in space group $\mathcal{G}$ only in conjunction with non-vanishing non-primitive translations $\mathcal{T}_R \neq 0$. The point group $\mathcal{G}_0$ of a non-symmorphic space group $\mathcal{G}$ is not contained as a subgroup in space group $\mathcal{G}$,

$$\mathcal{G}_0 \not\subset \mathcal{G}. \tag{11}$$

Formally one may write

$$\mathcal{G} = \mathbf{T} \odot \left\{ \{R, \mathcal{T}_R\} \,\middle|\, R \in \mathcal{G}_0 \right\}, \tag{12}$$

the symbol $\odot$ in this formula denoting the product (not a direct or semi-direct product) of the invariant subgroup $\mathbf{T}$ of pure primitive translations with the set $\left\{ \{R, \mathcal{T}_R\} \,\middle|\, R \in \mathcal{G}_0 \right\}$ of fundamental elements $\{R, \mathcal{T}_R\}$ of space group $\mathcal{G}$.

Since in the succeeding sections the quantities of prime interest will be scalar fields rather than mere vectors in configuration space, a mathematical framework is needed within which space group symmetries can be handled in a manner that is suited to function spaces. To this end, linear and unitary scalar transformation operators $P_n(\{R, \mathcal{T}_R + \mathbf{t}\})$ may be defined in the linear space of scalar functions $f(\mathbf{r}_1, \mathbf{r}_2, \cdots, \mathbf{r}_n)$ of variables $\mathbf{r}_i \in |\mathrm{R}^3$, $i = 1, 2, \cdots, n$, by means of

$$\left[P_n(\{R, \mathcal{T}_R + \mathbf{t}\}) \, f \right](\mathbf{r}_1, \mathbf{r}_2, \cdots, \mathbf{r}_n)$$
$$= f\left(\{R, \mathcal{T}_R + \mathbf{t}\}^{-1} \mathbf{r}_1, \{R, \mathcal{T}_R + \mathbf{t}\}^{-1} \mathbf{r}_2, \cdots, \{R, \mathcal{T}_R + \mathbf{t}\}^{-1} \mathbf{r}_n \right). \tag{13}$$

This definition of operators $P_n(\{R, \mathcal{T}_R + \mathbf{t}\})$ ensures that the product of two such operators is the scalar transformation operator of the product of the corresponding elements of space group $\mathcal{G}$,

$$P_n(\{R_1, \mathcal{T}_{R_1} + \mathbf{t}_1\}) \, P_n(\{R_2, \mathcal{T}_{R_2} + \mathbf{t}_2\}) = P_n(\{R_1, \mathcal{T}_{R_1} + \mathbf{t}_1\}\{R_2, \mathcal{T}_{R_2} + \mathbf{t}_2\}). \tag{14}$$

It is useful to rephrase definition (13) in somewhat more explicit wording. As emphasized by the braces on the left-hand side of this relation, prescription (13) defines, for any given $\{R, \mathcal{T}_R + \mathbf{t}\} \in \mathcal{G}$, an operator $P_n(\{R, \mathcal{T}_R + \mathbf{t}\})$ such that for any given scalar function f and all $(\mathbf{r}_1, \mathbf{r}_2, \cdots, \mathbf{r}_n) \in |\mathrm{R}^{3n}$ the value of the transformed function $P_n(\{R, \mathcal{T}_R + \mathbf{t}\}) \, f$ at $(\mathbf{r}_1, \mathbf{r}_2, \cdots, \mathbf{r}_n)$ will be given by the value of function f at $\left(\{R, \mathcal{T}_R + \mathbf{t}\}^{-1} \mathbf{r}_1, \{R, \mathcal{T}_R + \mathbf{t}\}^{-1} \mathbf{r}_2, \cdots, \{R, \mathcal{T}_R + \mathbf{t}\}^{-1} \mathbf{r}_n \right)$. For briefness of notation the braces will be henceforth omitted in all expressions involving functions transformed according to the rule (13).

With operator multiplication defined in the usual manner, the set $\left\{ P_n(\{R, \mathcal{T}_R+\mathbf{t}\}) \,\middle|\, \{R, \mathcal{T}_R+\mathbf{t}\} \in \mathcal{G} \right\}$ of scalar transformation operators, as defined in (13), becomes a group that is isomorphic to space group $\mathcal{G}$ – for short, although less precise, that is group $\mathcal{G}$ itself. Seen from a somewhat different point of view, the group of operators $P_n(\{R, \mathcal{T}_R + \mathbf{t}\})$ is isomorphic to the diagonal subgroup of that group of configuration space transformations in $3n-$dimensional space $|\mathbf{R}^{3n}$ that comprises the elements of the n-fold direct product of space group $\mathcal{G}$ with itself, this direct product being in turn a subgroup of all configuration space transformations in $|\mathbf{R}^{3n}$.

3 Correlated Many-Body Wave Functions

In the following, the invariance properties of various explicit forms of correlated many-body wave functions that are widely employed to study the ground state of quantum fluids and solids will be explored. The ensuing analysis focuses on quantum phases whose spatial invariance group is one of the 230 crystallographic space groups.

A widely adopted form for the correlated ground state wave function Ψ_0 of a Bose system of N identical particles is given by the Feenberg form (Feenberg 1969)

$$
\begin{aligned}
\Psi(\mathbf{r}_1, \mathbf{r}_2, \cdots, \mathbf{r}_N) &= \Psi_0(\mathbf{r}_1, \mathbf{r}_2, \cdots, \mathbf{r}_N) \\
&= \exp\Bigg\{ \frac{1}{2}\Bigg[\sum_{i=1}^{N} t(\mathbf{r}_i) + \sum_{i<j=1}^{N} u^{(2)}(\mathbf{r}_i, \mathbf{r}_j) \\
&\qquad\qquad + \sum_{i<j<k=1}^{N} u^{(3)}(\mathbf{r}_i, \mathbf{r}_j, \mathbf{r}_k) + \cdots \Bigg] \Bigg\},
\end{aligned} \qquad (15)
$$

incorporating one-body terms $t(\mathbf{r})$, two-body terms $u^{(2)}(\mathbf{r}_1, \mathbf{r}_2)$, three-body terms $u^{(3)}(\mathbf{r}_1, \mathbf{r}_2, \mathbf{r}_3)$ etc. The system is assumed to be confined to a normalization volume V made out of a finite number L of primitive unit cells with periodic boundary conditions imposed in this finite normalization volume, which in effect is tantamount to replacing the infinite group of primitive lattice translations by a finite group of order L. The form (15) must be symmetric under exchange of any two particles, entailing that the various n-body terms must be symmetric under exchange of any two sets of arguments $\mathbf{r}_1, \mathbf{r}_2, \mathbf{r}_3$ etc.

For description of the ground state of a Bose quantum solid a wave function of the type (15) is often supplemented with a product of cell orbitals $\varphi(\mathbf{r})$ localizing the particles to the sites $\mathbf{A}_i$, $i = 1, 2, \cdots, N$, of the crystal lattice, yielding the Nosanow form (Nosanow 1964)

$$
\Psi(\mathbf{r}_1, \mathbf{r}_2, \cdots, \mathbf{r}_N) = \prod_{i=1}^{N} \varphi(\mathbf{r}_i - \mathbf{A}_i)\, \Psi_0(\mathbf{r}_1, \mathbf{r}_2, \cdots, \mathbf{r}_N) \qquad (16)
$$

for the ground state wave function Ψ of a Bose quantum crystal.

Another type of wave function that has been applied successfully to model the ground state of a many-body Bose system is furnished by a shadow wave function (Vitiello, Runge, Chester, and Kalos 1990)

$$\Psi_s(\mathbf{r}_1, \mathbf{r}_2, \cdots, \mathbf{r}_N) = \Psi(\mathbf{r}_1, \mathbf{r}_2, \cdots, \mathbf{r}_N) \int \prod_{i=1}^{N} \varphi_{ps}(\mathbf{r}_i - \mathbf{s}_i)$$
$$\times \Phi(\mathbf{s}_1, \mathbf{s}_2, \cdots, \mathbf{s}_N) \, ds_1 ds_2 \cdots ds_N , \quad (17)$$

wherein the shadow part Φ is chosen to be of the form (15) and wherein space group symmetries allow Ψ to be of the type (15) or (16). The various analogous quantities involved in Ψ_0 and Φ need of course not be identical. The particle-shadow correlation factor $\varphi_{ps}(\mathbf{r})$ binds the real particles to the auxiliary shadow variables $\mathbf{s}_i$, $i = 1, 2, \cdots, N$.

Suitable wave functions for the ground state of a many-Fermion system may be obtained by multiplying Bosonic wave functions of the type (15), (16), or (17) with a Slater determinant of single-particle orbitals. In the many-Fermion wave functions arrived at in this manner, spin-spin correlations and the spin- and state dependencies of the spatial correlations, that in the general case will be present, are neglected. Such a many-Fermion ground state wave function makes sense only, if spin-spin interactions and spin- and state dependencies of spatial interactions in the Hamiltonian may also be neglected.

For a fully polarized state with all particle spins pointing in the same direction, the Slater determinant $\Phi_0(\mathbf{r}_1, \mathbf{r}_2, \cdots, \mathbf{r}_N)$ is assumed to be of the form

$$\Phi_0(\mathbf{r}_1, \mathbf{r}_2, \cdots, \mathbf{r}_N) = \det\{\psi_{k(i)}(\mathbf{r}_i) \, | \, 1 \le i \le N\}. \quad (18)$$

For an unpolarized state of total spin zero with half of the particles in a spin-up and another half of the particles in a spin-down state, the Slater determinant reads

$$\Phi_0(\mathbf{r}_1, \mathbf{r}_2, \cdots, \mathbf{r}_N; s_1, s_2, \cdots, s_N) = \det\{\psi_{k(i)}(\mathbf{r}_i) \, \chi_{S(i)}(s_i) \, | \, 1 \le i \le N\}, \quad (19)$$

wherein s_i, $i = 1, 2, \cdots, N$, with values $s_i = \pm 1$, are the spin variables of the particles and where $S(i) = 1$ for a spin-up and $S(i) = -1$ for a spin-down state. The spin orbitals $\chi_{s'}(s)$ are given by $\chi_{s'}(s) = \delta_{s's}$.

We assume the spatial single-particle orbitals $\psi_k(\mathbf{r})$, simply enumerated by $k = 1, 2, \cdots, N$ in (18) and by $k = 1, 2, \cdots, N/2$ in (19), to have been obtained as the lowest-lying solutions of an effective one-body Schrödinger equation of the form

$$\left\{ -\frac{\hbar^2}{2m} \nabla^2 + v(\mathbf{r}) \right\} \psi_k(\mathbf{r}) = \varepsilon_k \, \psi_k(\mathbf{r}) \quad (20)$$

with an effective single-particle potential $v(\mathbf{r})$ that is an invariant of space group $\mathcal{G}$, i. e.

$$P(\{R, \mathcal{T}_R + \mathbf{t}\}) \, v(\mathbf{r}) = v(\mathbf{r}), \quad \forall \{R, \mathcal{T}_R + \mathbf{t}\} \in \mathcal{G}. \quad (21)$$

The single-particle orbitals $\psi_k(\mathbf{r})$ may be grouped into sets $\{\psi_\kappa^{\alpha k p}(\mathbf{r}) \mid 1 \leq \kappa \leq d(\mathbf{k}, \mathrm{p})\}$ of basis functions, to the same eigenvalue $\varepsilon_\alpha(\mathbf{k}, \mathrm{p})$ of (20), of invariant, irreducible carrier spaces of unitary, irreducible representations (IR) $\Gamma_\mathcal{G}^{\mathbf{k}\mathrm{p}}$ of dimension $d(\mathbf{k}, \mathrm{p})$ of space group $\mathcal{G}$. A symbol $\Gamma_\mathcal{G}^{\mathbf{k}\mathrm{p}}$ denotes the IR of $\mathcal{G}$ obtained from vector $\mathbf{k}$ out of an asymmetric unit of the Brillouin zone and from loaded, irreducible representation $\hat{\Gamma}_{\mathcal{G}_0(\mathbf{k})}^{\mathrm{p}}$ of little point group $\mathcal{G}_0(\mathbf{k})$ of vector $\mathbf{k}$. The label α is used to distinguish invariant, irreducible subspaces of identical symmetry $\Gamma_\mathcal{G}^{\mathbf{k}\mathrm{p}}$ from each other. The basis functions $\psi_\kappa^{\alpha k p}$, $1 \leq \kappa \leq d(\mathbf{k}, \mathrm{p})$, transform among each other as the rows of IR $\Gamma_\mathcal{G}^{\mathbf{k}\mathrm{p}}$ of $\mathcal{G}$,

$$P(\{R, \mathcal{T}_R + \mathbf{t}\}) \psi_\kappa^{\alpha k p}(\mathbf{r}) = \sum_{\nu=1}^{d(\mathbf{k}, \mathrm{p})} \Gamma_\mathcal{G}^{\mathbf{k}\mathrm{p}} (\{R, \mathcal{T}_R + \mathbf{t}\})_{\nu \kappa} \, \psi_\nu^{\alpha k p}(\mathbf{r}). \tag{22}$$

For all types of wave functions discussed above the spatial n-body densities

$$\rho^{(n)}(\mathbf{r}_1', \mathbf{r}_2', \cdots, \mathbf{r}_n')$$

$$= \frac{\left\langle \Psi \left| \sum_{i_1 \neq i_2 \neq \cdots \neq i_n = 1}^{N} \delta(\mathbf{r}_1' - \mathbf{r}_{i_1}) \, \delta(\mathbf{r}_2' - \mathbf{r}_{i_2}) \cdots \delta(\mathbf{r}_n' - \mathbf{r}_{i_n}) \right| \Psi \right\rangle}{\langle \Psi | \Psi \rangle} \tag{23}$$

will be invariants of space group $\mathcal{G}$, that is

$$P_n(\{R, \mathcal{T}_R + \mathbf{t}\}) \, \rho^{(n)}(\mathbf{r}_1', \mathbf{r}_2', \cdots, \mathbf{r}_n') = \rho^{(n)}(\mathbf{r}_1', \mathbf{r}_2', \cdots, \mathbf{r}_n'), \quad \forall \{R, \mathcal{T}_R + \mathbf{t}\} \in \mathcal{G}, \tag{24}$$

and thus will correctly reflect the space group symmetries of the invariance group of the particular material under investigation, provided the various quantities entering the construction of a ground state wave function Ψ satisfy the following conditions.

1. The bosonic real-particle part Ψ_0, defined in (15), is an invariant of $\mathcal{G}$,

$$P_N(\{R, \mathcal{T}_R + \mathbf{t}\}) \, \Psi_0(\mathbf{r}_1, \mathbf{r}_2, \cdots, \mathbf{r}_N) = \Psi_0(\mathbf{r}_1, \mathbf{r}_2, \cdots, \mathbf{r}_N), \quad \forall \{R, \mathcal{T}_R + \mathbf{t}\} \in \mathcal{G}. \tag{25}$$

This property must be postulated also for the analogous shadow-correlation part $\Phi(\mathbf{s}_1, \mathbf{s}_2, \cdots, \mathbf{s}_N)$ of (17).

2. The cell orbital function φ in (16) is an invariant of the point group $\mathcal{G}_0$ of space group $\mathcal{G}$,

$$P(R) \varphi(\mathbf{r}) = \varphi(R^{-1}\mathbf{r}) = \varphi(\mathbf{r}), \quad \forall R \in \mathcal{G}_0. \tag{26}$$

The same property must hold for the particle-shadow correlation factor $\varphi_{ps}(\mathbf{r})$ in wave function (17). As far as property (26) of $\varphi(\mathbf{r})$ and $\varphi_{ps}(\mathbf{r})$ is concerned, no distinction needs to be made between symmorphic and non-symmorphic space groups.

3. With an orbital $\psi_\kappa^{\alpha \mathbf{k} \mathbf{p}}(\mathbf{r})$ occupied, by one particle in case of a Slater determinant of the form (18) and by two particles of opposite spin in case of a Slater determinant of the form (19), also all its partners to the same eigenvalue (same α) and IR $\Gamma_{\mathcal{G}}^{\mathbf{k} \mathbf{p}}$ of $\mathcal{G}$ are occupied by the same number of particles. Systematic degeneracies in a Slater determinant (18) or (19) are completely filled.

To proof the invariance property (24) in the case of Fermions, one must observe that a Slater determinant of the form (18) or (19) transforms according to

$$P_N\left(\{R, \mathcal{T}_R + \mathbf{t}\}\right) \Phi_0(\mathbf{r}_1, \mathbf{r}_2, \cdots, \mathbf{r}_N; s_1, s_2, \cdots, s_N)$$
$$= \left(\det U^T\right) \Phi_0(\mathbf{r}_1, \mathbf{r}_2, \cdots, \mathbf{r}_N; s_1, s_2, \cdots, s_N), \tag{27}$$

wherein U turns out to be a unitary matrix with matrices $\Gamma_{\mathcal{G}}^{\mathbf{k} \mathbf{p}}(\{R, \mathcal{T}_R + \mathbf{t}\})$ along its diagonal and zero entries else. A Slater determinant Φ_0 of the form (18) or (19) thus transforms into itself times a phase factor of absolute value 1. The absolute square of such a Slater determinant is an invariant of $\mathcal{G}$. The expectation value (23) includes integration (summation) over the spin coordinates s_i of all particles.

4 One-Body and Two-Body Functions

In this section the consequences of invariance property (25) for the one-body and two-body pseudopotentials $t(\mathbf{r})$ and $u^{(2)}(\mathbf{r}_1, \mathbf{r}_2)$ in (15) will be investigated. With a correspondingly larger amount of effort methods similar to the ones outlined here and applied to space group P6$_3$/mmc in the next section may be applied also to three-body functions such as the three-body pseudopotential $u^{(3)}(\mathbf{r}_1, \mathbf{r}_2, \mathbf{r}_3)$ in (15). The procedure will yield $t(\mathbf{r})$ and $u^{(2)}(\mathbf{r}_1, \mathbf{r}_2)$ in the form of orthogonal series in complete sets of symmetrized basis functions of variables in $|\mathbb{R}^3$ and $|\mathbb{R}^6$, respectively. Although cast in terms of $t(\mathbf{r})$ and $u^{(2)}(\mathbf{r}_1, \mathbf{r}_2)$, the following analysis applies in like vein also to the spatial one-body density $\rho(\mathbf{r})$ and the spatial two-body density $\rho^{(2)}(\mathbf{r}_1, \mathbf{r}_2)$ (better: $\rho^{(2)}(\mathbf{r}_1, \mathbf{r}_2) - \rho(\mathbf{r}_1)\rho(\mathbf{r}_2) = \rho(\mathbf{r}_1)\rho(\mathbf{r}_2)\left(g(\mathbf{r}_1, \mathbf{r}_2) - 1\right)$ for reasons of integrability) or, for that matter, to any other one-body or two-body quantity (such as, for example, $g(\mathbf{r}_1, \mathbf{r}_2) - 1$). The same is true for the analogous one-shadow and shadow-shadow pseudopotentials in a wave function of type (17).

A constant term in $t(\mathbf{r})$ may always be omitted, as such a term gives rise to an irrelevant factor in a many-body wave function that cancels in all expectation values, which must be taken with respect to normalized wave functions. One may assume $u^{(2)}(\mathbf{r}_1, \mathbf{r}_2)$ to be of the form

$$u^{(2)}(\mathbf{r}_1, \mathbf{r}_2) = u_0\left(|\mathbf{r}_1 - \mathbf{r}_2|\right) + u_{12}(\mathbf{r}_1, \mathbf{r}_2), \tag{28}$$

wherein the spherically symmetric function $u_0\left(|\mathbf{r}_1 - \mathbf{r}_2|\right)$ is supposed to embody the hard-core repulsion, when two particles approach each other very closely. In the following we will deal with $u_{12}(\mathbf{r}_1, \mathbf{r}_2)$ in (28), which too may contain a spherically symmetric part. The Fourier transform of $u_{12}(\mathbf{r}_1, \mathbf{r}_2)$ is assumed to exist.

The two-body function $u_{12}(\mathbf{r}_1, \mathbf{r}_2)$ is most efficiently dealt with by moving from particle variables $\mathbf{r}_1$ and $\mathbf{r}_2$ to center of mass

$$\mathbf{S} = \tfrac{1}{2}(\mathbf{r}_1 + \mathbf{r}_2) \tag{29}$$

and relative coordinate

$$\mathbf{r} = \mathbf{r}_1 - \mathbf{r}_2 \tag{30}$$

variables. Upon defining

$$u(\mathbf{S}, \mathbf{r}) = u_{12}\left(\mathbf{S} + \tfrac{1}{2}\mathbf{r}, \mathbf{S} - \tfrac{1}{2}\mathbf{r}\right), \tag{31}$$

the scalar transformation operators $P_2(\{R, \mathcal{T}_R + \mathbf{t}\})$ assume the convenient form

$$P_2(\{R, \mathcal{T}_R + \mathbf{t}\})\, u_{12}(\mathbf{r}_1, \mathbf{r}_2) = P_{\mathbf{S}}(\{R, \mathcal{T}_R + \mathbf{t}\})\, P_{\mathbf{r}}(R)\, u(\mathbf{S}, \mathbf{r})$$
$$= u\left(\{R, \mathcal{T}_R + \mathbf{t}\}^{-1}\mathbf{S}, R^{-1}\mathbf{r}\right), \tag{32}$$

providing a partial separation of translational and rotational degrees of freedom. The invariance properties of $t(\mathbf{r})$ and, in the $\mathbf{S}, \mathbf{r}$ variables, of $u(\mathbf{S}, \mathbf{r})$ become

$$P(\{R, \mathcal{T}_R + \mathbf{t}\})\, t(\mathbf{r}) = t(\{R, \mathcal{T}_R + \mathbf{t}\}^{-1}\mathbf{r}) = t(\mathbf{r}), \quad \forall\, \{R, \mathcal{T}_R + \mathbf{t}\} \in \mathcal{G} \tag{33}$$

and

$$P_{\mathbf{S}}(\{R, \mathcal{T}_R + \mathbf{t}\})\, P_{\mathbf{r}}(R)\, u(\mathbf{S}, \mathbf{r}) = u(\mathbf{S}, \mathbf{r}), \quad \forall\, \{R, \mathcal{T}_R + \mathbf{t}\} \in \mathcal{G}. \tag{34}$$

In eqs. (32) and (34) operator $P_{\mathbf{S}}(\{R, \mathcal{T}_R + \mathbf{t}\})$ is acting on only the center of mass variable $\mathbf{S}$, whereas the relative coordinate vector $\mathbf{r}$ is affected by only the rotational part R of space group element $\{R, \mathcal{T}_R + \mathbf{t}\}$. The factorization of function space operators $P_2(\{R, \mathcal{T}_R + \mathbf{t}\})$, as given by (32), enables one to take advantage of the linearity of configuration space rotations R, defined in (3).

4.1 Translational Symmetries

Because of

$$P(\{1, \mathbf{t}\})\, t(\mathbf{r}) = t(\mathbf{r} - \mathbf{t}) = t(\mathbf{r}), \quad \forall\, \{1, \mathbf{t}\} \in \mathbf{T}, \tag{35}$$

and

$$P_{\mathbf{S}}(\{1, \mathbf{t}\})\, u(\mathbf{S}, \mathbf{r}) = u(\mathbf{S} - \mathbf{t}, \mathbf{r}) = u(\mathbf{S}, \mathbf{r}), \quad \forall\, \{1, \mathbf{t}\} \in \mathbf{T}, \forall\, \mathbf{r} \in |\mathbb{R}^3, \tag{36}$$

the one-body and two-body pseudopotentials $t(\mathbf{r})$ and $u(\mathbf{S}, \mathbf{r})$ possess the periodicity of the Bravais lattice associated with space group $\mathcal{G}$ and thus may be readily expanded in Fourier series of plane waves $e^{i\mathbf{K}\mathbf{r}}$ and $e^{i\mathbf{K}\mathbf{S}}$, wherein vectors $\mathbf{K}$ are the reciprocal lattice vectors. The Fourier expansions read

$$t(\mathbf{r}) = \sum_{\mathbf{K}} t(\mathbf{K})\, e^{i\mathbf{K}\mathbf{r}}, \tag{37}$$

$$t(\mathbf{K}) = \frac{1}{\Omega} \int_{\Omega} t(\mathbf{r}) \, e^{-i\,\mathbf{Kr}} \, d\mathbf{r} \,, \tag{38}$$

and

$$u(\mathbf{S}, \mathbf{r}) = \sum_{\mathbf{K}} e^{i\,\mathbf{KS}} \, u(\mathbf{r}; \mathbf{K}) \,, \tag{39}$$

$$u(\mathbf{r}; \mathbf{K}) = \frac{1}{\Omega} \int_{\Omega} u(\mathbf{S}, \mathbf{r}) \, e^{-i\,\mathbf{KS}} \, d\mathbf{S} \,, \tag{40}$$

wherein Ω is the volume of a primitive unit cell and wherein the integrations in (38) and (40) extend over a primitive unit cell.

4.2 Rotational Symmetries

Exploiting invariance properties (33) and (34) furthermore for all rotations $R \in \mathcal{G}_0$ in the point group $\mathcal{G}_0$ of space group $\mathcal{G}$, formulas (38) and (40) may be employed to yield

$$t(R\mathbf{K}) = e^{-i\,(R\mathbf{K})\,\mathcal{T}_R} \, t(\mathbf{K}) \,, \quad \forall R \in \mathcal{G}_0 \,, \tag{41}$$

$$u(\mathbf{r}; R\mathbf{K}) = e^{-i\,(R\mathbf{K})\,\mathcal{T}_R} \, P_r(R) \, u(\mathbf{r}; \mathbf{K}) \,, \quad \forall R \in \mathcal{G}_0 \,. \tag{42}$$

The significance of relations (41) and (42) may be made most transparent by analyzing these results in terms of the invariance group $\mathcal{G}_0(\mathbf{K})$ of reciprocal lattice vector $\mathbf{K}$ and a concomitant decomposition of point group $\mathcal{G}_0$ into left cosets w. r. t. $\mathcal{G}_0(\mathbf{K})$.

Let

$$\mathcal{G}_0(\mathbf{K}) = \left\{ R' \in \mathcal{G}_0 \,\middle|\, R'\mathbf{K} = \mathbf{K} \right\} \subseteq \mathcal{G}_0 \tag{43}$$

be the point group of reciprocal lattice vector $\mathbf{K}$. For $\mathbf{K}=0$ point group $\mathcal{G}_0(\mathbf{K} = 0)$ will be identical to $\mathcal{G}_0$ itself. For a non-vanishing reciprocal lattice vector $\mathbf{K}$ the corresponding point group $\mathcal{G}_0(\mathbf{K})$ will be a subgroup of continuous point group $\infty\mathrm{m}(\mathbf{K}) = C_{\infty v}$, comprising all proper rotations about axis $\mathbf{K}$ and all reflections across mirrors containing this axis. Note that the definition of $\mathcal{G}_0(\mathbf{K})$ for a reciprocal lattice vector $\mathbf{K}$ differs from the definition of the little point group $\mathcal{G}_0(\mathbf{k}) = \left\{ R' \in \mathcal{G}_0 \,\middle|\, R'\mathbf{k} = \mathbf{k} + \mathbf{b}; \, \mathbf{b} \text{ a reciprocal lattice vector} \right\}$ for a vector $\mathbf{k}$ of the Brillouin zone. The latter definition of $\mathcal{G}_0(\mathbf{k})$ must be invoked for construction of the irreducible representations $\Gamma_{\mathcal{G}}^{\mathrm{kp}}$ of a crystallographic space group $\mathcal{G}$. Here, one is dealing in the end with only the totally symmetric IR of $\mathcal{G}$ with $\mathbf{k} = 0$. Let furthermore

$$\mathcal{G}_0 = \sum_{j=1}^{M(\mathbf{K})} R_j \, \mathcal{G}_0(\mathbf{K}) \tag{44}$$

be the decomposition of the point group $\mathcal{G}_0$ of space group $\mathcal{G}$ into left cosets $R_j \, \mathcal{G}_0(\mathbf{K})$ with respect to $\mathcal{G}_0(\mathbf{K})$. The set $\left\{ R_j \in \mathcal{G}_0 \,\middle|\, 1 \le j \le M(\mathbf{K}) \right\}$ is a set of representatives of expansion (44) of $\mathcal{G}_0$ into left cosets w.r.t. $\mathcal{G}_0(\mathbf{K})$, wherein

$$M(\mathbf{K}) = g_0 / g_0(\mathbf{K}) \tag{45}$$

with g_0 being the order of $\mathcal{G}_0$ and $g_0(\mathbf{K})$ the order of invariance group $\mathcal{G}_0(\mathbf{K})$ of reciprocal lattice vector $\mathbf{K}$. For convenience one may always choose $R_1 = E$ = identity element of $\mathcal{G}_0$. It may be shown that all results derived below are independent of the choice of representatives in the decomposition (44). The star $\mathbf{K}^*$ of reciprocal lattice vector $\mathbf{K}$ is given by

$$\mathbf{K}^* = \{R_j\mathbf{K} \,|\, 1 \leq j \leq M(\mathbf{K})\}. \tag{46}$$

In conjunction with (12) the expansion formula (44) provides an efficient means of handling the in general non-Abelian structure of space groups.

Equations (41) and (42) furnish relationships between, respectively, the $R_j\mathbf{K}$−components $t(R_j\mathbf{K})$ and $u(\mathbf{r}; R_j\mathbf{K})$ belonging to the symmetry-equi-valent reciprocal lattice vectors $R_j\mathbf{K}$, $1 \leq j \leq M(\mathbf{K})$, in a given star $\mathbf{K}^*$. The expansion coefficients $t(R_j\mathbf{K})$ and the components $u(\mathbf{r}; R_j\mathbf{K})$ are not independent of each other but may all be obtained from $t(\mathbf{K})$ and $u(\mathbf{r}; \mathbf{K})$, respectively, by means of

$$t(R_j\mathbf{K}) = e^{-i(R_j\mathbf{K})\mathcal{T}_j}\, t(\mathbf{K}), \;\; 1 \leq j \leq M(\mathbf{K}), \tag{47}$$

$$u(\mathbf{r}; R_j\mathbf{K}) = e^{-i(R_j\mathbf{K})\mathcal{T}_j}\, P_{\mathbf{r}}(R_j)\, u(\mathbf{r}; \mathbf{K}), \;\; 1 \leq j \leq M(\mathbf{K}), \tag{48}$$

permitting to cast expansions (37) and (39) in the form

$$t(\mathbf{r}) = \sum_{\mathbf{K} \in AS} t(\mathbf{K}) \sum_{j=1}^{M(\mathbf{K})} e^{i(R_j\mathbf{K})(\mathbf{r} - \mathcal{T}_j)}, \tag{49}$$

$$u(\mathbf{S}, \mathbf{r}) = \sum_{\mathbf{K} \in AS} \sum_{j=1}^{M(\mathbf{K})} e^{i(R_j\mathbf{K})(\mathbf{s} - \mathcal{T}_j)}\, P_{\mathbf{r}}(R_j)\, u(\mathbf{r}; \mathbf{K}), \tag{50}$$

wherein $\sum_{\mathbf{K} \in AS}$ runs through all reciprocal lattice vectors that are in different stars. Results are independent of the choice made for an asymmetric unit AS of symmetry-inequivalent reciprocal lattice vectors.

Beyond leading to expressions (49) and (50), equations (41) and (42), due to

$$t(\mathbf{K}) = e^{i\mathbf{K}\mathcal{T}_{R'}}\, t(\mathbf{K}), \quad \forall\, R' \in \mathcal{G}_0(\mathbf{K}), \tag{51}$$

$$P_{\mathbf{r}}(R')\, u(\mathbf{r}; \mathbf{K}) = e^{i\mathbf{K}\mathcal{T}_{R'}}\, u(\mathbf{r}; \mathbf{K}), \quad \forall\, R' \in \mathcal{G}_0(\mathbf{K}), \tag{52}$$

impose also that, for any given $\mathbf{K} \in AS$, functions $t(\mathbf{K})$ and $u(\mathbf{r}; \mathbf{K})$, the former being just a constant function, transform under the rotations $R' \in \mathcal{G}_0(\mathbf{K})$ as the (one and only) row(s) of that one-dimensional, unitary, irreducible representation $\Gamma^{\mathcal{T}}_{\mathcal{G}_0(\mathbf{K})}$ of point group $\mathcal{G}_0(\mathbf{K})$ that is furnished by the mapping

$$\Gamma^{\mathcal{T}}_{\mathcal{G}_0(\mathbf{K})}(R') = e^{i\mathbf{K}\mathcal{T}_{R'}}, \quad \forall\, R' \in \mathcal{G}_0(\mathbf{K}). \tag{53}$$

Because of a mere constant being an invariant of any group, we find that $t(\mathbf{K})$ must vanish whenever $e^{i\mathbf{K}\mathcal{T}_{R'}} \neq 1$ for anysome $R' \in \mathcal{G}_0(\mathbf{K})$. With

$$n_1(\mathbf{K}) = \frac{1}{g_0(\mathbf{K})} \sum_{R' \in \mathcal{G}_0(\mathbf{K})} e^{i\mathbf{K}\mathcal{T}_{R'}} \;\; (= \text{either } 0 \text{ or } 1) \tag{54}$$

being the number of times the identity representation of $\mathcal{G}_0(\mathbf{K})$ occurs in the representation of $\mathcal{G}_0(\mathbf{K})$ given by the mapping (53), expansion (49) may be written as

$$t(\mathbf{r}) = \sum_{\mathbf{K} \in AS} t(\mathbf{K}) \left\{ n_1(\mathbf{K}) \sum_{j=1}^{M(\mathbf{K})} e^{i\left(R_j \mathbf{K}\right)\left(\mathbf{r}-\mathcal{T}_j\right)} \right\}, \tag{55}$$

which is an expansion of $t(\mathbf{r})$ in a complete set of symmetrized, with respect to the trivial identity representation $\Gamma_{\mathcal{G}}^{01}$ of the full space group $\mathcal{G}$, plane waves $n_1(\mathbf{K}) \sum_{j=1}^{M(\mathbf{K})} e^{i\left(R_j \mathbf{K}\right)\left(\mathbf{r}-\mathcal{T}_j\right)}$. The character projection operator $\mathcal{P}_{\mathcal{G}}^{01} = (g_0 L)^{-1}$ $\sum_t \sum_{R \in \mathcal{G}_0} P(\{R, \mathcal{T}_R + \mathbf{t}\})$ of the totally symmetric IR of $\mathcal{G}$ applied to $e^{i\mathbf{K}\mathbf{r}}$ yields the same result except for a factor $1/M(\mathbf{K})$ that may be absorbed into $t(\mathbf{K})$. Here, L is the total number of primitive unit cells in a finite normalization volume V. The symmetry-adapted basis functions are not normalized.

Whereas $t(\mathbf{K})$ must vanish, whenever the representation (53) of $\mathcal{G}_0(\mathbf{K})$ does not coincide with the identity representation, the two-body components $u(\mathbf{r}; \mathbf{K})$ need not be zero for such a reciprocal lattice vector $\mathbf{K} \in AS$. However, equation (52) establishes that in such a case $u(\mathbf{r}; \mathbf{K})$ does not transform as the identity representation of $\mathcal{G}_0(\mathbf{K})$. The latter is the case for $n_1(\mathbf{K}) = 1$. For a symmorphic space group we have $n_1(\mathbf{K}) = 1$ for all $\mathbf{K}$.

Equations (39), (48), (50), and (52) particularly clearly reveal the structure of $u(\mathbf{S}, \mathbf{r})$ that is resulting from the underlying space group symmetry. For every star $\mathbf{K}^*$ there is merely a single independent component, $u(\mathbf{r}; \mathbf{K})$, the symmetry of which w. r. t. $\mathcal{G}_0(\mathbf{K})$ is given by the transformation properties (52). The component associated with the rotated vector $R_j \mathbf{K} \in \mathbf{K}^*$ is obtained by rotating function $u(\mathbf{r}; \mathbf{K})$ and embellishing the result, $P_{\mathbf{r}}(R_j) u(\mathbf{r}; \mathbf{K})$, with the appropriate phase factor $e^{-i(R_j\mathbf{K})\mathcal{T}_j}$.

4.3 Exchange Symmetry

The non-space group exchange symmetry imposes on $u(\mathbf{r}; \mathbf{K})$ the additional constraint

$$u(-\mathbf{r}; \mathbf{K}) = u(\mathbf{r}; \mathbf{K}), \tag{56}$$

that is $u(\mathbf{r}; \mathbf{K})$ must be an even function of its argument $\mathbf{r}$, regardless of whether space group $\mathcal{G}$ is centrosymmetric or not.

Exchange symmetry thus has the effect of adding a center of inversion to point group $\mathcal{G}_0(\mathbf{K})$, thereby producing the Laue class $L(\mathbf{K})$ of point group $\mathcal{G}_0(\mathbf{K})$,

$$\begin{aligned} L(\mathbf{K}) &= \{E, I\} \otimes \mathcal{G}_0(\mathbf{K}) = \{R', IR' \mid R' \in \mathcal{G}_0(\mathbf{K})\} \\ &= \bar{1} \otimes \mathcal{G}_0(\mathbf{K}) = C_i \otimes \mathcal{G}_0(\mathbf{K}) \subseteq L(0). \end{aligned} \tag{57}$$

Here, the identity element is denoted by E and the inversion through the origin by I. The Laue class $L(\mathbf{K})$ is the direct product of the two commuting groups $\{E, I\} = \bar{1} = C_i$ and $\mathcal{G}_0(\mathbf{K})$. It is a subgroup of the Laue class $L(0) = \bar{1} \otimes \mathcal{G}_0 \supseteq \mathcal{G}_0$ of the point group $\mathcal{G}_0$ of space group $\mathcal{G}$. In formula (57), point group

$\{E, I\}$ is given in international notation, $\bar{1}$, as well as in Schoenflies notation, C_i. In plane groups the inversion I across the origin is equivalent to the proper rotation about the origin through an angle of 180^0. Even when space group $\mathcal{G}$ is centrosymmetric itself, although, depending on the choice of origin, there need not be an inversion center at the origin of the coordinate frame of reference, the inversion symmetry of functions $u(\mathbf{r}; \mathbf{K})$ is caused not by space group but by exchange symmetry. This distinction becomes pointless for the $\mathbf{K} = 0$ component $u(\mathbf{r}; 0)$ in centrosymmetric space groups. In all other cases, however, the intrinsic eigensymmetry of $u(\mathbf{r}; \mathbf{K})$ is higher than space group symmetry alone dictates. As a particular consequence, the $\mathbf{K} = 0$ term in an expansion of the form (48) allows to distinguish only between the eleven Laue classes and not between all of the 32 crystallographic point groups. Space group inversion symmetry, if present, will manifest itself in inversion symmetry of the stars $\mathbf{K}^*$, as defined in (46), throughout the whole reciprocal lattice.

Taking into account exchange symmetry, it follows from (52) that $u(\mathbf{r}; \mathbf{K})$ transforms under the (point group) elements of the respective Laue class $L(\mathbf{K})$ of reciprocal lattice vector $\mathbf{K}$ as that one-dimensional IR $\Gamma_{L(\mathbf{K})}^{\mathcal{T}}$ of group $L(\mathbf{K})$ that is provided by the mapping

$$\Gamma_{L(\mathbf{K})}^{\mathcal{T}}(R') = \Gamma_{L(\mathbf{K})}^{\mathcal{T}}(IR') = \Gamma_{\mathcal{G}_0(\mathbf{K})}^{\mathcal{T}}(R') = e^{i \mathbf{K} \mathcal{T}_{R'}}, \quad \forall R' \in \mathcal{G}_0(\mathbf{K}). \tag{58}$$

For a centrosymmetric space group $\mathcal{G}$ representations $\Gamma_{\mathcal{G}_0(\mathbf{K})}^{\mathcal{T}}$ and $\Gamma_{L(\mathbf{K})}^{\mathcal{T}}$ are real for all $\mathbf{K}$. For a space group lacking centers of inversion these representations may or may not be real.

5   Application to Space Group P6$_3$/mmc (D$_{6h}^4$, No. 194)

In this section the techniques of the general group-theoretical formalism developed in the previous sections will be applied specifically to space group P6$_3$/mmc (D$_{6h}^4$ in Schoenflies notation, No. 194 in the *International Tables for Crystallography* (ITC) (Hahn 1995)). An important crystalline structure displaying the P6$_3$/mmc space group symmetry is the hexagonal close-packed (hcp) structure (structure **A3** in "Strukturbericht" notation, prototype: magnesium). The helium isotopes, for example, or, for that matter, also the alkaline-earth metals Be and Mg all exhibit hcp solid phases.

The point group $\mathcal{G}_0$ of space group $\mathcal{G} = $ P6$_3$/mmc turns out to be $\mathcal{G}_0 = $ 6/mmm (D$_{6h}$) and thus coincides with the (holohedral) point group of the hexagonal Bravais lattice. In an xyz cartesian frame of reference with the $z-$axis pointing upwards in the vertical direction, the $x-$axis to the right, and the $y-$axis being directed away from the viewer, basis vectors of the direct Bravais lattice are given by

$$\mathbf{a}_1 = \frac{a}{2}\left(-1, -\sqrt{3}, 0\right), \quad \mathbf{a}_2 = a\left(1, 0, 0\right), \quad \text{and} \quad \mathbf{a}_3 = c\left(0, 0, 1\right), \tag{59}$$

wherein a and c are the lattice constants. The corresponding basis vectors $\mathbf{b}_1$, $\mathbf{b}_2$, and $\mathbf{b}_3$ of the reciprocal lattice may be derived to be

$$\mathbf{b}_1 = \frac{2\pi}{a}\left(0, -\frac{2}{\sqrt{3}}, 0\right), \quad \mathbf{b}_2 = \frac{2\pi}{a}\left(1, -\frac{1}{\sqrt{3}}, 0\right), \quad \text{and} \quad \mathbf{b}_3 = \frac{2\pi}{c}(0,0,1).$$
(60)

The choice (59) of basis vectors for the hexagonal primitive unit cell agrees with the one that is conventionally adopted by crystallographers. It is in accord also with the choice of basis vectors found in ITC as well as with the one in Kovalev's tables (Kovalev 1993). With the choice of origin of the frame of reference as in ITC, applied consistently throughout this paper, the particles in the hcp structure are at the $2c$ sites $\mathbf{A}_1 + \mathbf{t}$ and $\mathbf{A}_2 + \mathbf{t}$, wherein $\mathbf{t}$ represents the lattice vectors of the hexagonal Bravais lattice. The vectors

$$\mathbf{A}_1 = \tfrac{1}{3}\,\mathbf{a}_1 + \tfrac{2}{3}\,\mathbf{a}_2 + \tfrac{1}{4}\,\mathbf{a}_3 \quad \text{and} \quad \mathbf{A}_2 = \tfrac{2}{3}\,\mathbf{a}_1 + \tfrac{1}{3}\,\mathbf{a}_2 + \tfrac{3}{4}\,\mathbf{a}_3$$
(61)

denote the positions of the two particles in the first primitive unit cell, originating in a space group inversion center. In the hcp structure, the inversion centers of space group P6$_3$/mmc are not at the particle positions, whose site symmetry is point group $\bar{6}$m2 (D$_{3h}$), but, when projected onto the $\mathbf{a}_1\mathbf{a}_2-$plane, shifted from these and, in the vertical direction, half-way between the horizontal planes of crystal lattice sites.

Table 1 provides a complete list of the essential symmetry operations $\{R, \mathcal{T}_R\}$, $R \in \mathcal{G}_0$, as introduced in the context surrounding (12) in Sec. 2, of space group P6$_3$/mmc. According to prescription (12), the entire space group may then be obtained by multiplying the fundamental elements with the pure translations $\{1, \mathbf{t}\}$ by Bravais lattice vectors $\mathbf{t}$. As we shall see, for the present purposes it suffices, however, to work with only the fundamental elements, since the translational symmetries of functions $t(\mathbf{r})$ and $u(\mathbf{S}, \mathbf{r})$ are taken into account already by representing them as Fourier series, given by the expansions (37) and (39).

The first column in Table 1 refers to the numbers under which the fundamental space group elements, given in the second column, may be found listed, below the space group diagram of P6$_3$/mmc, in ITC. In the second column, the essential symmetry operations of this space group are displayed in the international notation, which is the one that is used also in ITC. The hexagonal coordinates x_h, y_h, and z_h in Table 1 may be obtained as the projections of $\mathbf{r} = (x, y, z)$ along axes $\mathbf{a}_1$, $\mathbf{a}_2$, and $\mathbf{a}_3$. They are related to the cartesian variables x, y, and z by means of

$$x = y_h - \tfrac{1}{2}\,x_h, \quad y = -\tfrac{\sqrt{3}}{2}\,x_h, \quad \text{and} \quad z = z_h.$$
(62)

Note that in ITC as well as in Ref.(Kovalev 1993) it is the hexagonal variables x_h, y_h, and z_h that are denoted by the letters x, y, and z in contrast to this work, in which the latter represent cartesian coordinates and, for distinction from these, symbols x_h, y_h, and z_h are used instead to denote coordinates in the hexagonal frame of reference.

Although the international notation, in particular in conjunction with a space group diagram, allows one to fairly quickly establish the geometrical meaning of space group elements, it is far too cumbersome to be used effectively in group algebraic calculations. For these the symbols used in the *Point-Group Theory Tables* (PGTT) of S. L. Altmann and P. Herzig (Altmann and Herzig 1994) are adopted for denoting the (proper and improper) point group rotations. Adapted to this notation, the fundamental elements of space group $P6_3/mmc$ assume the form shown in the third column of Table 1. Clearly, the symbols $C_6^\pm$, $C_3^\pm$, and C_2 represent the proper rotations in the counter-clockwise sense about the principal z-axis through angles of, respectively, $\pm 60^0$, $\pm 120^0$, and 180^0. The identity element is denoted by E, the inversion through the origin by I, and the reflection across the horizontal $\mathbf{a}_1\mathbf{a}_2(xy)$-plane by σ_h. Referring to hexagonal coordinates, the pure rotations through an angle of 180^0 about axes $\mathbf{a}_2$, $\mathbf{a}_1 + \mathbf{a}_2$, $\mathbf{a}_1$, $2\mathbf{a}_1 + \mathbf{a}_2$, $\mathbf{a}_1 - \mathbf{a}_2$, and $\mathbf{a}_1 + 2\mathbf{a}_2$ are, in this order, denoted by C'_{2i} with $i = 1, 2, 3$ and by C''_{2i} with $i = 1, 2, 3$. Likewise, the symbols σ_{di} with $i = 1, 2, 3$ and σ_{vi} with $i = 1, 2, 3$ are used to denote, in the given order, the mirror reflections across the vertical planes perpendicular to $\mathbf{a}_2$, $\mathbf{a}_1 + \mathbf{a}_2$, $\mathbf{a}_1$, $2\mathbf{a}_1 + \mathbf{a}_2$, $\mathbf{a}_1 - \mathbf{a}_2$, and $\mathbf{a}_1 + 2\mathbf{a}_2$. The rotation-inversions $S_3^\mp$ and $S_6^\mp$ are defined by means of $S_3^\mp = IC_6^\pm$ and $S_6^\mp = IC_3^\pm$.

Space group $P6_3/mmc$ is non-symmorphic, the principal axis being a 6_3 screw axis, and possesses three vertical c-glide mirror planes. With the choice of origin as in ITC we have

$$\mathcal{T}_R = 0 \quad \text{for} \quad R = E, C_3^\pm, C'_{2i}, I, S_6^\mp, \sigma_{di} \tag{63}$$

and

$$\mathcal{T}_R = \mathcal{T} = \tfrac{1}{2}\mathbf{a}_3 \quad \text{for} \quad R = C_6^\pm, C_2, C''_{2i}, S_3^\mp, \sigma_h, \sigma_{vi}. \tag{64}$$

The set of essential symmetry operations of crystallographic space group $P6_3/mmc$, listed in Table 1, contains six truly non-symmorphic transformations, namely the three screw rotations $\{C_6^\pm, \mathcal{T}\}$ and $\{C_2, \mathcal{T}\}$, Nos. 6, 5, and 4 in Table 1, and the three vertical c-glide mirror reflections $\{\sigma_{vi}, \mathcal{T}\}$ with $i = 1, 2, 3$, Nos. 24, 22, and 23 in Table 1.

In the last column of Table 1 the essential elements of $P6_3/mmc$ are listed in the notation used in the space group tables compiled by O. V. Kovalev (Kovalev 1993). In these tables the point group operations are simply enumerated. The configuration space operators h_i, $1 \le i \le 24$, appearing in the last column of Table 1, are, of course, the H-system rotations given by Kovalev. As is the case also with most other space groups, the choice of origin in Kovalev's tables for $P6_3/mmc$ coincides with the one adopted in ITC for this space group.

A set of reciprocal lattice vectors representing the entity of inequivalent stars is given by

$$AS = \left\{ \mathbf{K} = n_1\mathbf{b}_1 + n_2\mathbf{b}_2 + n_3\mathbf{b}_3 \,\big|\, n_1, n_2, n_3 \in |\mathrm{N}_0\,;\, n_2 \le n_1 \right\}. \tag{65}$$

The reciprocal lattice vectors $\mathbf{K} \in AS$ may be classified according to their point groups $\mathcal{G}_0(\mathbf{K})$. Vectors whose point groups $\mathcal{G}_0(\mathbf{K})$ are identical may be classified further according to which IR of $\mathcal{G}_0(\mathbf{K})$ is produced by the mapping $\Gamma^{\mathcal{T}}_{\mathcal{G}_0(\mathbf{K})}$, defined in (53). The result of this analysis is shown in Table 2, where the

Table 1. The essential elements of hexagonal space group P6$_3$/mmc

ITC	ITC	PGTT	Kovalev
1	1	$\{E, 0\}$	$\{h_1, 0\}$
6	6^+ $\left(0,0,\frac{1}{2}\right)$ $0,0,z_h$	$\{C_6^+, \mathcal{T}\}$	$\{h_2, \mathcal{T}\}$
5	6^- $\left(0,0,\frac{1}{2}\right)$ $0,0,z_h$	$\{C_6^-, \mathcal{T}\}$	$\{h_6, \mathcal{T}\}$
2	3^+ $\quad 0,0,z_h$	$\{C_3^+, 0\}$	$\{h_3, 0\}$
3	3^- $\quad 0,0,z_h$	$\{C_3^-, 0\}$	$\{h_5, 0\}$
4	2 $\left(0,0,\frac{1}{2}\right)$ $0,0,z_h$	$\{C_2, \mathcal{T}\}$	$\{h_4, \mathcal{T}\}$
9	2 $\quad 0,y_h,0$	$\{C'_{21}, 0\}$	$\{h_7, 0\}$
7	2 $\quad x_h,x_h,0$	$\{C'_{22}, 0\}$	$\{h_{11}, 0\}$
8	2 $\quad x_h,0,0$	$\{C'_{23}, 0\}$	$\{h_9, 0\}$
12	2 $\quad 2x_h,x_h,\frac{1}{4}$	$\{C''_{21}, \mathcal{T}\}$	$\{h_{10}, \mathcal{T}\}$
10	2 $\quad x_h,\overline{x}_h,\frac{1}{4}$	$\{C''_{22}, \mathcal{T}\}$	$\{h_8, \mathcal{T}\}$
11	2 $\quad x_h,2x_h,\frac{1}{4}$	$\{C''_{23}, \mathcal{T}\}$	$\{h_{12}, \mathcal{T}\}$
13	$\overline{1}$ $\quad 0,0,0$	$\{I, 0\}$	$\{h_{13}, 0\}$
18	$\overline{6}^+$ $\quad 0,0,z_h$; $0,0,\frac{1}{4}$	$\{S_3^-, \mathcal{T}\}$	$\{h_{14}, \mathcal{T}\}$
17	$\overline{6}^-$ $\quad 0,0,z_h$; $0,0,\frac{1}{4}$	$\{S_3^+, \mathcal{T}\}$	$\{h_{18}, \mathcal{T}\}$
14	$\overline{3}^+$ $\quad 0,0,z_h$; $0,0,0$	$\{S_6^-, 0\}$	$\{h_{15}, 0\}$
15	$\overline{3}^-$ $\quad 0,0,z_h$; $0,0,0$	$\{S_6^+, 0\}$	$\{h_{17}, 0\}$
16	m $\quad x_h,y_h,\frac{1}{4}$	$\{\sigma_h, \mathcal{T}\}$	$\{h_{16}, \mathcal{T}\}$
21	m $\quad 2x_h,x_h,z_h$	$\{\sigma_{d1}, 0\}$	$\{h_{19}, 0\}$
19	m $\quad x_h,\overline{x}_h,z_h$	$\{\sigma_{d2}, 0\}$	$\{h_{23}, 0\}$
20	m $\quad x_h,2x_h,z_h$	$\{\sigma_{d3}, 0\}$	$\{h_{21}, 0\}$
24	c $\quad 0,y_h,z_h$	$\{\sigma_{v1}, \mathcal{T}\}$	$\{h_{22}, \mathcal{T}\}$
22	c $\quad x_h,x_h,z_h$	$\{\sigma_{v2}, \mathcal{T}\}$	$\{h_{20}, \mathcal{T}\}$
23	c $\quad x_h,0,z_h$	$\{\sigma_{v3}, \mathcal{T}\}$	$\{h_{24}, \mathcal{T}\}$

symmetry-inequivalent reciprocal lattice vectors $\mathbf{K}$ are classified according to invariance groups $\mathcal{G}_0(\mathbf{K})$, defined in (43), and irreducible representations $\Gamma^{\mathcal{T}}_{\mathcal{G}_0(\mathbf{K})}$ of $\mathcal{G}_0(\mathbf{K})$, furnished by the mapping (53), for space group P6$_3$/mmc. The first column in this table simply defines a label to distinguish the various types of reciprocal lattice vectors $\mathbf{K}$, given explicitly in the second column, that result from this classification scheme. In the third column, point group $\mathcal{G}_0(\mathbf{K})$ is given in international as well as, in parentheses, in Schoenflies notation with the point

Table 2. Classification of symmetry-inequivalent reciprocal lattice vectors $\mathbf{K} \in AS$

	$\mathbf{K}$	$\mathcal{G}_0(\mathbf{K})$	$n_1(\mathbf{K})$	$\Gamma^{\mathcal{T}}_{\mathcal{G}_0(\mathbf{K})}$
0	0	6/mmm (D$_{6h}$) $\{E, C_6^{\pm}, C_3^{\pm}, C_2,$ $C'_{2i}, C''_{2i}, I, S_3^{\mp},$ $S_6^{\mp}, \sigma_h, \sigma_{di}, \sigma_{vi}\}$	1	A_{1g}
1	$(2n_3+1)\mathbf{b}_3$ $n_3 \in |\mathbb{N}_0$	6mm (C$_{6v}$) $\{E, C_6^{\pm}, C_3^{\pm},$ $C_2, \sigma_{di}, \sigma_{vi}\}$	1	B_2
1'	$2n_3\mathbf{b}_3$ $n_3 \in |\mathbb{N}$	6mm (C$_{6v}$) $\{E, C_6^{\pm}, C_3^{\pm},$ $C_2, \sigma_{di}, \sigma_{vi}\}$	1	A_1
2	$n_1\mathbf{b}_1$ $n_1 \in |\mathbb{N}$	mm2 (C$_{2v}$) $\{E, C''_{21}, \sigma_{d1}, \sigma_h\}$	1	A_1
3	$n_1(\mathbf{b}_1+\mathbf{b}_2)$ $n_1 \in |\mathbb{N}$	mm2 (C$_{2v}$) $\{E, C'_{22}, \sigma_{v2}, \sigma_h\}$	1	A_1
4	$n_1\mathbf{b}_1 + n_2\mathbf{b}_2$ $n_1, n_2 \in |\mathbb{N}$ $n_2 < n_1$	m (C$_s$) $\{E, \sigma_h\}$	1	A'
5	$n_1\mathbf{b}_1 + n_3\mathbf{b}_3$ $n_1, n_3 \in |\mathbb{N}$	m (C$_s$) $\{E, \sigma_{d1}\}$	1	A'
6	$n_1(\mathbf{b}_1+\mathbf{b}_2)$ $+(2n_3+1)\mathbf{b}_3$ $n_1 \in |\mathbb{N}, n_3 \in |\mathbb{N}_0$	m (C$_s$) $\{E, \sigma_{v2}\}$	0	A''
6'	$n_1(\mathbf{b}_1+\mathbf{b}_2)$ $+2n_3\mathbf{b}_3$ $n_1, n_3 \in |\mathbb{N}$	m (C$_s$) $\{E, \sigma_{v2}\}$	1	A'
7	$n_1\mathbf{b}_1 + n_2\mathbf{b}_2$ $+n_3\mathbf{b}_3$ $n_1, n_2, n_3 \in |\mathbb{N}$ $n_2 < n_1$	1 (C$_1$) $\{E\}$	1	A

group rotations contained in $\mathcal{G}_0(\mathbf{K})$ listed in curly braces below the point group symbol. The integer $n_1(\mathbf{K})$, defined in (54), assumes the values entered in the fourth column. For $n_1(\mathbf{K}) = 1$, one-dimensional irreducible representation $\Gamma^{\mathcal{T}}_{\mathcal{G}_0(\mathbf{K})}$ of $\mathcal{G}_0(\mathbf{K})$, the symmetry label of which is given in the last column of this table, coincides with the trivial identity representation of $\mathcal{G}_0(\mathbf{K})$. Except for reciprocal lattice vectors of type 1 or 6, representation $\Gamma^{\mathcal{T}}_{\mathcal{G}_0(\mathbf{K})}$ of $\mathcal{G}_0(\mathbf{K})$, shown in the last column in Table 2, coincides with the identity representation of that group. The Fourier transform $t(\mathbf{K})$ of $t(\mathbf{r})$ vanishes for reciprocal lattice vectors of type 1 or 6. When applied to $t(\mathbf{r})$, the symmetrization procedure described in the last section yields the result

$$
\begin{aligned}
t(\mathbf{r}) = \sum_{n_3=0}^{\infty} \sum_{n_1=0}^{\infty} \sum_{n_2=0}^{n_1} t(n_1, n_2, n_3) & \Big\{ \cos\Big(\big[n_1 \mathbf{b}_1 + n_2 \mathbf{b}_2 + n_3 \mathbf{b}_3 \big] \mathbf{r} \Big) \\
& + \cos\Big(\big[(n_1 + n_2)\mathbf{b}_1 - n_1 \mathbf{b}_2 - n_3 \mathbf{b}_3 \big] \mathbf{r} \Big) \\
& + \cos\Big(\big[n_2 \mathbf{b}_1 - (n_1 + n_2)\mathbf{b}_2 + n_3 \mathbf{b}_3 \big] \mathbf{r} \Big) \\
& + \cos\Big(\big[(n_1 + n_2)\mathbf{b}_1 - n_2 \mathbf{b}_2 + n_3 \mathbf{b}_3 \big] \mathbf{r} \Big) + \cos\Big(\big[n_2 \mathbf{b}_1 + n_1 \mathbf{b}_2 - n_3 \mathbf{b}_3 \big] \mathbf{r} \Big) \\
& + \cos\Big(\big[n_1 \mathbf{b}_1 - (n_1 + n_2)\mathbf{b}_2 - n_3 \mathbf{b}_3 \big] \mathbf{r} \Big) \\
& + (-1)^{n_3} \Big[\cos\Big(\big[n_2 \mathbf{b}_1 - (n_1 + n_2)\mathbf{b}_2 - n_3 \mathbf{b}_3 \big] \mathbf{r} \Big) \\
& + \cos\Big(\big[(n_1 + n_2)\mathbf{b}_1 - n_1 \mathbf{b}_2 + n_3 \mathbf{b}_3 \big] \mathbf{r} \Big) + \cos\Big(\big[n_1 \mathbf{b}_1 + n_2 \mathbf{b}_2 - n_3 \mathbf{b}_3 \big] \mathbf{r} \Big) \\
& + \cos\Big(\big[(n_1 + n_2)\mathbf{b}_1 - n_2 \mathbf{b}_2 - n_3 \mathbf{b}_3 \big] \mathbf{r} \Big) + \cos\Big(\big[n_2 \mathbf{b}_1 + n_1 \mathbf{b}_2 + n_3 \mathbf{b}_3 \big] \mathbf{r} \Big) \\
& + \cos\Big(\big[n_1 \mathbf{b}_1 - (n_1 + n_2)\mathbf{b}_2 + n_3 \mathbf{b}_3 \big] \mathbf{r} \Big) \Big] \Big\}.
\end{aligned}
\tag{66}
$$

The expression in curly braces is the formula for the (unnormalized) w. r. t. $P6_3/mmc$ totally symmetric plane wave for a general reciprocal lattice vector $\mathbf{K} = n_1 \mathbf{b}_1 + n_2 \mathbf{b}_2 + n_3 \mathbf{b}_3$ of type 7 in Table 2. Formula (66) encompasses also all other cases, corresponding to high symmetry reciprocal lattice vectors, that are listed in Table 2. It simplifies for those $\mathbf{K}$ and, in particular, yields a vanishing result for reciprocal lattice vectors of type 1 or 6, which amounts to setting, for these latter types of $\mathbf{K}$−vectors, the Fourier transform $t(\mathbf{K}) = 0$ in the series (49). The factor $(-1)^{n_3}$, with which altogether six of the twelve cosine-terms have to be multiplied in (66), is a consequence of the non-symmorphic nature of space group $P6_3/mmc$.

Suitable sets of basis functions in which functions $u(\mathbf{r}; \mathbf{K})$ may be represented can be obtained from spherical harmonics

$$
Y_l^m \left(\frac{x}{r}, \frac{y}{r}, \frac{z}{r}; \frac{\mathbf{K}}{K} \right) = i^{m+|m|} P_l^{|m|} \left(\frac{\mathbf{K}\mathbf{r}}{Kr} \right) \left(\frac{R(\mathbf{K})_{j1}\, x_j + i\, R(\mathbf{K})_{j2}\, x_j}{\sqrt{r^2 - R(\mathbf{K})_{j3}\, R(\mathbf{K})_{l3}\, x_j\, x_l}} \right)^m,
\tag{67}
$$

wherein $K = |\mathbf{K}|$ and $r = |\mathbf{r}|$. Here, the phase convention adopted in Ref.(Altmann and Herzig 1994) is followed. For $\mathbf{K} \neq 0$, the real orthogonal 3×3 matrix $R(\mathbf{K})$ represents, in the xyz−frame, the transformation that sends the z−axis into the direction of $\mathbf{K}$ and the x−axis into the direction given by the outer product $\mathbf{b}_3 \times \mathbf{K}$, this vector being perpendicular to the vertical plane spanned by vectors $\mathbf{b}_3$ and $\mathbf{K}$. Transformation $R(\mathbf{K})$ thus produces a new, righthanded, cartesian $x'y'z'$−frame with the z'−axis parallel to $\mathbf{K}$ and the x'−axis, in the xy−plane, parallel to $\mathbf{b}_3 \times \mathbf{K}$. The $y'z'$−plane coincides with the $\mathbf{b}_3\mathbf{K}$−plane. In compliance with the usual convention in tensor calculus one has to sum over double indices in formula (67) wherever applicable. The same rule is to be applied throughout the remainder of this section. The associated Legendre functions in (67) read

$$P_l^{|m|}(\xi) = \sqrt{\frac{(2l+1)(l-|m|)!}{4\pi(l+|m|)!}} \frac{1}{2^l l!} \left(1-\xi^2\right)^{\frac{|m|}{2}} \frac{d^{l+|m|}}{d\xi^{l+|m|}} \left(\xi^2-1\right)^l. \tag{68}$$

For any given reciprocal lattice vector $\mathbf{K} = n_1\mathbf{b_1} + n_2\mathbf{b_2} + n_3\mathbf{b_3}$, matrix $R(\mathbf{K})$ in (67) may be constructed from the product of a counter-clockwise proper rotation about the $x-$axis through an angle γ given by

$$\cos\gamma = \frac{n_3}{\sqrt{\left(\frac{2c}{\sqrt{3}\,a}\right)^2 \left(n_1^2 + n_1 n_2 + n_2^2\right) + n_3^2}} \tag{69}$$

with a counter-clockwise proper rotation about the $z-$axis through an angle β determined by

$$\cos\beta = \frac{2n_1 + n_2}{2\sqrt{n_1^2 + n_1 n_2 + n_2^2}}. \tag{70}$$

The angle γ is the angle enclosed by vectors $\mathbf{b_3}$ and $\mathbf{K}$, the former of these vectors being parallel to the $z-$axis. Likewise, the angle β is the one that vector $\mathbf{b_1}$, parallel to the $-y-$direction, encloses with the projection of $\mathbf{K}$ onto the $\mathbf{b_1}\mathbf{b_2}(xy)-$plane. The result for $R(\mathbf{K})$ assumes the form

$$R(\mathbf{K}) = \begin{bmatrix} \cos\beta & -\sin\beta\,\cos\gamma & \sin\beta\,\sin\gamma \\ \sin\beta & \cos\beta\,\cos\gamma & -\cos\beta\,\sin\gamma \\ 0 & \sin\gamma & \cos\gamma \end{bmatrix}. \tag{71}$$

The matrix-valued function $R(\mathbf{K})$ depends, of course, on only the direction $\mathbf{K}/K$ of $\mathbf{K}$. For $\mathbf{K} = 0$ or for $\mathbf{K}$ parallel to $\mathbf{b_3}$, it becomes the 3×3 identity matrix. In these cases, one obviously has $\mathbf{Kr}/Kr \longrightarrow z/r$ for the argument of the associated Legendre functions in (67).

Formula (67) derives from the standard expression $Y_l^m(\mathbf{r}'/|\mathbf{r}'|) = i^{m+|m|}$ $\times P_l^{|m|}(\cos\theta')\,e^{i\,m\varphi'}$ for the spherical harmonics in the $x'y'z'-$frame, wherein θ' and φ' are the polar and azimuthal angles of a coordinate vector $\mathbf{r}'$ in this frame, by first casting the complex exponential in the form $\exp(\pm i\,|m|\varphi') = \eta'^{-|m|}\left(x'\pm iy'\right)^{|m|}$ with $\eta' = \left(x'^2+y'^2\right)^{\frac{1}{2}}$ and by inserting $\cos\theta' = z'/|\mathbf{r}'| = z'/r$. The expression (67) is then obtained by exploiting the fact that the position vectors $\mathbf{r}' = (x',y',z')$ and $\mathbf{r} = (x,y,z)$ of a point in space, when referred to, respectively, the new $x'y'z'-$ and the original $xyz-$frame, are related by means of $\mathbf{r}' = R^T(\mathbf{K})\,\mathbf{r}$, wherein $R^T(\mathbf{K})$ denotes the transpose of $R(\mathbf{K})$. This latter relation corresponds in fact to the passive interpretation, specialized to cartesian frames, of a point group rotation introduced in its active sense in (3) of Sec. 2. Cast in the form (67), the spherical harmonics are polynomial expressions in x_i/r with $i = 1, 2, 3$ corresponding to x/r, y/r, and z/r. The square root term in the denominator in expression (67) cancels against the corresponding one in the enumerator of the Legendre functions (68). Whence, replacing x_i by x_i/r for all i, the square root terms could equally well be simply omitted in these equations. Provided appropriate circumspection is being exercised in this regard, the same applies, in an analoguous fashion, for such or similar square root denominators or factors in other formulas of this section too.

Symmetrized basis functions for representing two-body quantities could be constructed as well on basis of the "usual" spherical harmonics $Y_l^m(\mathbf{r}/r)$ in the xyz−frame or, for that matter, from harmonics $Y_l^m(\mathbf{r}/r; \mathbf{g})$ with $\mathbf{g}$ being any direction in space. For any given $\mathbf{K}$ and value of l, identical function spaces will be spanned by maximal sets of linearly independent, w. r. t. IR $\Gamma_{L(\mathbf{K})}^{\mathcal{T}}$ of Laue class $L(\mathbf{K})$ symmetrized functions projected out from either set of $2l + 1$ harmonics, $Y_l^m(\mathbf{r}/r)$ or else $Y_l^m(R^{-1}(\mathbf{K})\,\mathbf{r}/r)$ with $0 \le |m| \le l$ in both cases. The function space thus obtained necessarily will be a $L(\mathbf{K})$−invariant subspace of symmetry $\Gamma_{L(\mathbf{K})}^{\mathcal{T}}$ of some dimension d between 0 and $2l + 1$. For $d > 1$, this w. r. t. $L(\mathbf{K})$ reducible subspace will in turn be composed of d one-dimensional subspaces of symmetry $\Gamma_{L(\mathbf{K})}^{\mathcal{T}}$. The rationale underlying the choice (67) is that it permits a rather straightforward classification of symmetry-adapted basis functions according to the irreducible representations of continuous point group ∞om ($C_{\infty v}$), comprising all proper rotations about a given axis and all reflections across mirrors containing that axis. Evidently, this purpose singles out the $\mathbf{K}$−direction. Irreducible representations of ∞om in turn can be related to irreducible representations of the Euclidean group of all translations, rotations, and products thereof in three-dimensional space. In this respect, the choice of the $x'-$ and $y'-$axes is irrelevant. It is, however, nonetheless of some conceptual and technical advantage to have all axes of the $x'y'z'-$frame oriented along high-symmetry directions of the respective Laue class $L(\mathbf{K})$. One can easily convince oneself that the choice of the $x'y'z'-$frames, described in some detail in the context of equations (69)–(71), satisfies these requirements for all special $\mathbf{K}$−vectors, belonging to one of the types 0–6 listed in Table 2. As far as the $x'-$ and $y'-$axes are concerned, the latter symmetry arguments cannot be invoked anymore in case of the general $\mathbf{K}$−vectors of type 7 in Table 2. For non-vanishing $\mathbf{K}$, only subgroup $\mathcal{G}_0(\mathbf{K})$ of Laue class $L(\mathbf{K})$ is determined by space group symmetry, whereas the remaining half of elements in $L(\mathbf{K})$, making up for subset $I\,\mathcal{G}_0(\mathbf{K}) = \{IR' \mid R' \in \mathcal{G}_0(\mathbf{K})\} \subset L(\mathbf{K})$, is generated by exchange symmetry.

Final results for symmetry-adapted basis functions, in which the $\mathbf{K}$−components

$$u(\mathbf{S}, \mathbf{r}; \mathbf{K}) = \sum_{j=1}^{M(\mathbf{K})} e^{i\,(R_j\mathbf{K})\,(\mathbf{s}-\mathcal{T}_j)}\,P_{\mathbf{r}}(R_j)\,u(\mathbf{r}; \mathbf{K}), \qquad (72)$$

invariants of $\mathcal{G}$ for all $\mathbf{K}$ (cf. (50)), of $u(\mathbf{S}, \mathbf{r})$ can be represented, may be conveniently expressed in terms of the real functions

$$P_{\mathbf{r}}(R) \left\langle \frac{\mathbf{r}}{r} \middle| 1, n; \frac{\mathbf{K}}{K} \right\rangle = \left(r^2 - R(\mathbf{K})_{j3}\, R(\mathbf{K})_{l3}\, R_{ij}\, R_{ml}\, x_i\, x_m \right)^{-\frac{n}{2}}$$

$$\times \sum_{k=0}^{\left[\frac{n}{2}\right]} (-1)^k \binom{n}{2k} \left[R(\mathbf{K})_{j1}\, R_{lj}\, x_l \right]^{n-2k} \left[R(\mathbf{K})_{j2}\, R_{lj}\, x_l \right]^{2k} \tag{73}$$

and

$$P_{\mathbf{r}}(R) \left\langle \frac{\mathbf{r}}{r} \middle| 2, n; \frac{\mathbf{K}}{K} \right\rangle = \left(r^2 - R(\mathbf{K})_{j3}\, R(\mathbf{K})_{l3}\, R_{ij}\, R_{ml}\, x_i\, x_m \right)^{-\frac{n}{2}}$$

$$\times \sum_{k=0}^{\left[\frac{n-1}{2}\right]} (-1)^k \binom{n}{2k+1} \left[R(\mathbf{K})_{j1}\, R_{lj}\, x_l \right]^{n-(2k+1)}$$

$$\times \left[R(\mathbf{K})_{j2}\, R_{lj}\, x_l \right]^{2k+1}, \tag{74}$$

wherein $n \in |\mathrm{N}_0$ in (73) and $n \in |\mathrm{N}$ in (74). With regard to the reality of functions (73) and (74), it should be emphasized that the symmetrization procedure, as outlined in Sec. 4, is, with a minimum chance of subtle errors, best carried through to its very end exclusively on the set of complex numbers. A real basis may then be obtained by taking the real and imaginary part of the, in general up to this point, complex basis functions.

Making use of PGTT, the $\mathbf{K} = 0$ term $u(\mathbf{r}; 0)$ may be worked out to be of the form

$$u(\mathbf{r}; 0) = \sum_{l=0}^{\infty} \sum_{k=0}^{2} \sum_{m=0}^{l} R_{2(3l+k)}^{6m}(r; 0)\, P_{2(3l+k)}^{6m}(\cos\theta)\, \cos(6m\varphi). \tag{75}$$

Owing to the 6_3 screw axis and the vertical c-glide mirror planes, function $u(\mathbf{r}; n_3\mathbf{b}_3)$ must be a basis function of IR B_2 of crystallographic point group 6mm (C_{6v}) for n_3 odd, whereas for n_3 ($\neq 0$) even it must be a basis function of the totally symmetric IR A_1 of 6mm, the latter situation being identical to the one encountered for all $n_3\mathbf{b}_3$, $n_3 \in |\mathrm{N}$, in symmorphic space group P6/mmm. One winds up with

$$u(\mathbf{r}; (2n_3+1)\mathbf{b}_3) = \sum_{n=0}^{\infty} \sum_{l=3n+2}^{\infty} R_{2l}^{6n+3}(r; (2n_3+1)\mathbf{b}_3)\, P_{2l}^{6n+3}(\cos\theta)$$

$$\times \sin((6n+3)\varphi), \tag{76}$$

$$u(\mathbf{r}; 2n_3\mathbf{b}_3) = \sum_{n=1}^{\infty} \sum_{l=3n}^{\infty} R_{2l}^{6n}(r; 2n_3\mathbf{b}_3)\, P_{2l}^{6n}(\cos\theta)\, \cos(6n\varphi)$$

$$+ \sum_{l=0}^{\infty} R_{2l}^0 (r; 2n_3 \mathbf{b}_3) \, P_{2l}^0 (\cos\theta) \,, \tag{77}$$

wherein $n_3 \in |N_0$ in (76) and $n_3 \in |N$ in (77). Equations (75)–(77) leave us with radial functions $R_{2(3l+k)}^{6m}(r; 0)$, $R_{2l}^{6n+3}(r; (2n_3 + 1)\mathbf{b}_3)$, $R_{2l}^0(r; 2n_3\mathbf{b}_3)$, and $R_{2l}^{6n}(r; 2n_3\mathbf{b}_3)$ that may be expanded further in suitable radial basis functions such as, for example, sperical Bessel functions $j_{2l}(qr)$ with $q \in |R_0^+$ or the ones used in Ref.(Moroni, Fantoni, and Senatore 1995) in a variational Monte Carlo optimization of two- and three-body correlation terms in isotropic ^{4}He and ^{3}He quantum liquids.

For a general reciprocal lattice vector of type 7 in Table 2, the expansion of $u(\mathbf{S}, \mathbf{r}; \mathbf{K})$ may be cast in the form

$$u(\mathbf{S}, \mathbf{r}; \mathbf{K}_7) = \sum_{n=0}^{\infty} \sum_{l=\left[\frac{n+1}{2}\right]}^{\infty} \left\langle r \,\Big|\, 1, n, 2l; \mathbf{K}_7 \right\rangle \left\langle \mathbf{S}, \frac{\mathbf{r}}{r} \,\Big|\, 1, n, 2l; \mathbf{K}_7 \right\rangle$$

$$+ \sum_{n=1}^{\infty} \sum_{l=\left[\frac{n+1}{2}\right]}^{\infty} \left\langle r \,\Big|\, 2, n, 2l; \mathbf{K}_7 \right\rangle \left\langle \mathbf{S}, \frac{\mathbf{r}}{r} \,\Big|\, 2, n, 2l; \mathbf{K}_7 \right\rangle \tag{78}$$

with radial coefficient functions $\langle r \,|\, \alpha, n, 2l; \mathbf{K}_7 \rangle$ and with $\alpha = 1, 2$. The symmetry-adapted basis functions $\left\langle \mathbf{S}, \frac{\mathbf{r}}{r} \,\big|\, \alpha, n, 2l; \mathbf{K}_7 \right\rangle$ may be written as

$$\left\langle \mathbf{S}, \frac{\mathbf{r}}{r} \,\Big|\, \alpha, n, 2l; \mathbf{K}_7 \right\rangle = \sum_{R \in \mathcal{S}^+} \cos\big((R\mathbf{K}_7)\mathbf{S}\big) \, P_{2l}^n\left(\frac{(R\mathbf{K}_7)\mathbf{r}}{K_7 r}\right)$$

$$\times P_{\mathbf{r}}(R) \left\langle \frac{\mathbf{r}}{r} \,\Big|\, \alpha, n; \frac{\mathbf{K}_7}{K_7} \right\rangle$$

$$+ (-1)^{n_3} \sum_{R \in \mathcal{S}^-} \cos\big((R\mathbf{K}_7)\mathbf{S}\big) \, P_{2l}^n\left(\frac{(R\mathbf{K}_7)\mathbf{r}}{K_7 r}\right) P_{\mathbf{r}}(R) \left\langle \frac{\mathbf{r}}{r} \,\Big|\, \alpha, n; \frac{\mathbf{K}_7}{K_7} \right\rangle, \tag{79}$$

wherein the first sum extends over the six rotations R in the set

$$\mathcal{S}^+ = \left\{ E, C_3^{\pm}, \sigma_{di} \right\}, \tag{80}$$

and the second over the six rotations R in the set

$$\mathcal{S}^- = \left\{ C_6^{\pm}, C_2, \sigma_{vi} \right\}. \tag{81}$$

Expressions similar to the ones given for the cases just disussed hold for the other types of reciprocal lattice vectors listed in Table 2. For reciprocal lattice vectors of type 2 or 3 the pertinent result reads

$$u(\mathbf{S}, \mathbf{r}; \mathbf{K}_{2,3}) = \sum_{n=0}^{\infty} \sum_{l=n}^{\infty} R_{2l}^{2n}(r; \mathbf{K}_{2,3}) \sum_{R \in \left\{ E, C_3^{\pm} \right\}} \cos\big((R\mathbf{K}_{2,3})\mathbf{S}\big)$$

$$\times P_{2l}^{2n}\left(\frac{(R\mathbf{K}_{2,3})\mathbf{r}}{K_{2,3}\,r}\right) P_{\mathbf{r}}(R) \left\langle \frac{\mathbf{r}}{r}\,\middle|\,1,2n;\frac{\mathbf{K}_{2,3}}{K_{2,3}}\right\rangle, \quad (82)$$

whereas for **K** of type 4 it assumes the form

$$u(\mathbf{S},\mathbf{r};\mathbf{K}_4) = \sum_{n=0}^{\infty}\sum_{l=\left[\frac{n+1}{2}\right]}^{\infty} R_{2l}^n(r;\mathbf{K}_4) \sum_{R\in\mathcal{S}^+} \cos\big((R\mathbf{K}_4)\mathbf{S}\big)\, P_{2l}^n\left(\frac{(R\mathbf{K}_4)\mathbf{r}}{K_4\,r}\right)$$

$$\times P_{\mathbf{r}}(R)\left\langle \frac{\mathbf{r}}{r}\,\middle|\,1,n;\frac{\mathbf{K}_4}{K_4}\right\rangle. \quad (83)$$

Likewise, one obtains for **K** of type 5 or 6' the expression

$$u(\mathbf{S},\mathbf{r};\mathbf{K}_{5,6'}) = \sum_{n=0}^{\infty}\left\{\sum_{l=n}^{\infty}\left\langle r\,\middle|\,1,2n,2l;\mathbf{K}_{5,6'}\right\rangle\left\langle \mathbf{S},\frac{\mathbf{r}}{r}\,\middle|\,1,2n,2l;\mathbf{K}_{5,6'}\right\rangle\right.$$

$$\left.+\sum_{l=n+1}^{\infty}\left\langle r\,\middle|\,2,2n+1,2l;\mathbf{K}_{5,6'}\right\rangle\left\langle \mathbf{S},\frac{\mathbf{r}}{r}\,\middle|\,2,2n+1,2l;\mathbf{K}_{5,6'}\right\rangle\right\} \quad (84)$$

and for **K** of type 6

$$u(\mathbf{S},\mathbf{r};\mathbf{K}_6) = \sum_{n=1}^{\infty}\sum_{l=n}^{\infty}\left\{\left\langle r\,\middle|\,2,2n,2l;\mathbf{K}_6\right\rangle\left\langle \mathbf{S},\frac{\mathbf{r}}{r}\,\middle|\,2,2n,2l;\mathbf{K}_6\right\rangle\right.$$

$$\left.+\left\langle r\,\middle|\,1,2n-1,2l;\mathbf{K}_6\right\rangle\left\langle \mathbf{S},\frac{\mathbf{r}}{r}\,\middle|\,1,2n-1,2l;\mathbf{K}_6\right\rangle\right\}. \quad (85)$$

The remaining quantities in eqs. (84) and (85) for **K** of type 5, 6', or 6 may be cast in the form

$$\left\langle \mathbf{S},\frac{\mathbf{r}}{r}\,\middle|\,\alpha,m,2l;\mathbf{K}_{5,6',6}\right\rangle = \sum_{R\in\{E,C_3^{\pm}\}} \cos\big((R\mathbf{K}_{5,6',6})\mathbf{S}\big)\, P_{2l}^m\left(\frac{(R\mathbf{K}_{5,6',6})\mathbf{r}}{K_{5,6',6}\,r}\right)$$

$$\times P_{\mathbf{r}}(R)\left\langle \frac{\mathbf{r}}{r}\,\middle|\,\alpha,m;\frac{\mathbf{K}_{5,6',6}}{K_{5,6',6}}\right\rangle$$

$$+(-1)^{n_3}\sum_{R\in\{C_6^{\pm},C_2\}} \cos\big((R\mathbf{K}_{5,6',6})\mathbf{S}\big)\, P_{2l}^m\left(\frac{(R\mathbf{K}_{5,6',6})\mathbf{r}}{K_{5,6',6}\,r}\right)$$

$$\times P_{\mathbf{r}}(R)\left\langle \frac{\mathbf{r}}{r}\,\middle|\,\alpha,m;\frac{\mathbf{K}_{5,6',6}}{K_{5,6',6}}\right\rangle. \quad (86)$$

Acknowledgement

This work is supported by the Graduiertenkolleg *Klassifizierung von Phasen-umwandlungen kristalliner Stoffe aufgrund struktureller und physikalischer Anomalien (Classification of phase transitions in crystalline materials due to structural and physical anomalies)* at the University of Cologne and by the European Union Human Capital and Mobility Scheme Network *Microscopic quantum many-body theory: applications to traditional and novel forms of matter.*

References

Altmann, S.L. and Herzig, P.: *Point-Group Theory Tables* (Clarendon Press, Oxford, 1994).

Alonso, J.A. and March, N.H.: *Electrons in Metals and Alloys* (Academic Press, London, 1989).

Bishop, R.F., Parkinson, J.B., and Xian, Y.: J. Phys. Condens. Matter **5**, 9169 (1993).

Bradley, C.J. and Cracknell, A.P.: *The Mathematical Theory of Symmetry in Solids* (Oxford U.P., Oxford, 1972).

Burns, G. and Glazer, A.M.: *Space Groups for Solid State Scientists* (Academic Press, Boston, 1990).

Campbell, C.E., Kürten, K.E., Ristig, M.L., and Senger, G.: Phys. Rev. B **30**, 3728 (1984).

Campbell, C.E., Clements, B.E., Krotscheck, E., and Saarela, M.: Phys. Rev. B, in press.

Chin, S.A. and Krotscheck, E.: Phys. Rev. B **45**, 852 (1992).

Clements, C.E, Krotscheck, E., and Tymczak, C.J.: Phys. Rev. B **53**, 12253 (1996).

Cornwell, J.F.: *Group Theory in Physics*, Vol. I (Academic Press, London, 1984).

Feenberg, E.: *Theory of Quantum Fluids* (Academic Press, New York, 1969).

Fulde, P.: *Electron Correlations in Molecules and Solids* (Springer-Verlag, Berlin, 1991).

Gernoth, K.A., Clark, J.W., Senger, G., and Ristig, M.L.: Phys. Rev. B **49**, 15836 (1994).

Gernoth, K.A., Clark, J.W., and Ristig, M.L.: in *Condensed Matter Theories*, Vol. 10, edited by Casas, M. et al. (Nova Science Publishers, Commack, NY, 1995).

Gernoth, K.A. and Clark, J.W.: J. Low Temp. Phys. **96**, 153 (1994).

Guirao, A., Centelles, M., Barranco, M., Pi, M., Polls, A., and Viñas, X.: J. Phys. Condens. Matter **4**, 667 (1992).

International Tables for Crystallography, Vol. A, 4th edition, edited by Hahn, Th. (Kluwer Academic Publishers, Dordrecht, 1995).

Krotscheck, E., Qian, G.-X., and Kohn, W.: Phys. Rev. **31**, 4545 (1985).

Krotscheck, E.: Phys. Rev. B **31**, 4258 (1985b).

Krotscheck, E.: Phys. Rev. B **31**, 4267 (1985).

Krotscheck, E.: in *Recent Progress in Many-Body Theories*, Vol. 2, edited by Avishai, Y. (Plenum Press, New York, 1990).

Kovalev, O.V.: *Representations of the Crystallographic Space Groups*, edited by Stokes, H.T. and Hatch, D.M. (Gordon and Breach Science Publishers, Switzerland, 1993).

March, N.H.: *Electron Density Theory of Atoms and Molecules* (Academic Press, London, 1992).

Moroni, S., Fantoni, S., and Senatore, G.: Europhys. Lett. **30**, 93 (1995).

Nosanow, L.H.: Phys. Rev. Lett. **13**, 270 (1964).

Pederiva, F., Ferrante A., Fantoni, S., and Reatto, L.: Phys. Rev. B **52**, 7564 (1995).

Pederiva, F., Vitiello, S.A., Gernoth, K.A., Fantoni, S., and Reatto, L.: Phys. Rev. B **53**, 15129 (1996).

Pederiva, F., Ferrante, A., Fantoni, S., and Reatto, L.: Phys. Rev. Lett. **72**, 2589 (1994).

Ristig, M.L. and Kim, J.W.: Phys. Rev. B **40**, 6665 (1996).

Saarela, M. and Suominen, J.: in *Condensed Matter Theories*, Vol. 4, edited by Keller, J. (Plenum Press, New York, 1989).

Saarela, M.: in *Recent Progress in Many-Body Theories*, Vol. 4, edited by Schachinger, E., Mitter, H., and Sormann, H. (Plenum Press, New York, 1995).

Saarela, M., Pietiläinen P., and Kallio, A.: Phys. Rev. B **27**, 231 (1983).

Senger, G., Ristig, M.L., Kürten, K.E., and Campbell, C.E.: Phys. Rev. B **33**, 7256 (1986).

Stringari, S. and Treiner, J.: Phys. Rev. B **36**, 8369 (1987).

Szybisz, L. and Ristig, M.L.: Phys. Rev. B **40**, 4391 (1989).

Tolédano, J.C. and Tolédano, P.: *The Landau Theory of Phase Transitions* (World Scientific, Singapore, 1987).

Vitiello, S.A. and Schmidt, K.E.: Phys. Rev. B **46**, 5442 (1992).

Vitiello, S.A., Runge, K.J., Chester, G.V., and Kalos, M.H.: Phys. Rev. B **42**, 228 (1990).

Wagner, M. and Ceperley, D.M.: J. Low Temp. Phys. **94**, 185 (1994).

Hamiltonian Lattice Calculations on Gauge and Chiral Meson Field Theories

Siu A. Chin

Center for Theoretical Physics, Department of Physics,
Texas A&M University, College Station, TX 77843, U.S.A.

Abstract. I review Hamiltonian lattice calculations on SU(N) gauge theories and present some recent results on chiral meson field theories. A common theme in both is the study of collective excitations, corresponding to glueball masses in gauge theories and vector meson masses in chiral meson field theories. In the case of the non-linear sigma model, I show that the variational mass gap of the ω-like collective mode vanishes as one approaches the chiral phase transition.

1 Introduction

In view of the growing interest in the study of Hamiltonian lattice gauge theories by various semi-analytic methods (see other contributions to this workshop), it might be useful to summarize some scattered Monte Carlo results in one place for future comparisons. I will therefore begin with a personal summary of known Hamiltonian lattice results, focusing mostly on glueball masses. I will then show that variational sum rules developed from these studies can be directly applied to the study of chiral lattice field theories, such as the non-linear sigma model. The latter can be used to study the equation of state of pions (Bunatian and Wambach, 1994), or masses of vector mesons at high temperature (Brown and Rho, 1991). Both are topics of current interest in relativistic heavy-ion collisions. Also, a latticized chiral field theory is essentially a quantized spin (or rotor) system. As such, it provides a novel variation to the theme of this workshop.

A Hamiltonian lattice field theory is very similar to other many-body systems such as liquid helium or finite nuclei. It has the advantage of having only very smooth potentials, such as the cosine function. Its unique difficulty is that its ground state, the vacuum, is not of interest and only excited states are physical. The basic problem is therefore to determine the system's collective excitations. In a pure gauge theory, these excitation energies are just glueball masses. In conventional many-body systems, one begins with a mean-field description of the ground state (Hartree-Fock), and then studies collective excitations as small amplitude oscillations about the mean-field (RPA). In lattice gauge theories, a mean-field theory of the fundamental link variable is not even gauge invariant. Thus one cannot even begin to apply conventional many-body wisdom to Hamiltonian lattice gauge theories.

In the absence of a detailed theory of gauge-invariant collective excitations, sum rules (Bohigas *et al.*, 1979) can be exploited to extract glueball masses.

The required matrix elements can be evaluated by the Monte Carlo method. This is the approach used here. The Monte Carlo method of directly simulating the dynamics of SU(N) link variables is a powerful and versatile way of solving lattice gauge theories. It can be applied to any gauge group at any dimension. Some semi-analytical methods are hampered by the need to know recoupling coefficients, which are unknown or unfamiliar beyond SU(2).

In this work, I will recall some salient features of Hamiltonian lattice gauge theories in section 2, emphasizing their connection with conventional many-body problems. Section 3 describes the basic variational sum rule approach for determining glueball masses. Section 4 summarizes results on the ground state energy, the scaling behavior of glueball masses and the continuum limit of various mass ratios. Details about Monte Carlo methods have been given elsewhere (Chin, 1986) and will not be repeated here. Section 5 describes the latticization of the non-linear sigma model. Section 6 gives results on its ground state energy and section 7 studies the mass gap of the ω-like collective excitation.

2 Hamiltonian Lattice Gauge Theory

A lattice gauge theory is an *integral* (Yang, 1974) formulation of a gauge-invariant theory with finite degrees of freedom. Instead of formulating the theory in terms of the field tensor, which involve *differentials* of the gauge field, one develops a theory in terms of the gauge phase, which is a path-dependent *integral* of the gauge potential:

$$U(x,y) = P \exp\left[-ig \int_x^y dx_\mu T^a A_\mu^a(x)\right]. \tag{1}$$

The minimum gauge phase between two lattice sites, $U(\mathbf{x}, \mathbf{x} + a\mathbf{e}_k) \equiv U_k(\mathbf{x})$, is regarded as a fundamental degree of freedom. Conjugate to it is the generalized electric field operator defined by the commutation relation,

$$[E_{l'}^a, U_l] = \delta_{l'l} T^a U_l, \tag{2}$$

where l, l' are link labels. Stoke's Theorem in the form

$$\exp\left[\int_{\partial\Omega} \mathbf{A} \cdot d\mathbf{s}\right] = \exp\left[\int_{\Omega} \mathbf{B} \cdot d\mathbf{a}\right], \tag{3}$$

where Ω is the encircled area, suggests that the product of $U_k(\mathbf{x})$'s around an elementary loop is related to the **B** field. To see that this is indeed the case, one notes that in the limit of small lattice spacing a, (1) is approximately

$$U_k(\mathbf{x}) \simeq \exp\left[-iagA_k(\mathbf{x})\right], \tag{4}$$

where $A = T^a A^a$. Hence the product of four $U_k(\mathbf{x})'$s encircling an elemental square, a plaquette, is

$$
\begin{aligned}
U_p &= U_1(\mathbf{x}) U_2(\mathbf{x} + a\mathbf{e}_1) U_1^{-1}(\mathbf{x} + a\mathbf{e}_2) U_2^{-1}(\mathbf{x}), \\
&\simeq \exp\left[-iag A_1\right] \exp\left[-iag(A_2 + a\partial_\mu A_2)\right] \exp\left[iag(A_1 + a\partial_\nu A_1)\right] \exp\left[iag A_2\right], \\
&\simeq \exp\left[-ia^2 g B_3(\mathbf{x}) + O(a^4)\right].
\end{aligned}
\tag{5}
$$

Similarly, plaquettes in the other two planes give B_1^a and B_2^a. The lattice analog of the continuum SU(N) gauge theory Hamiltonian,

$$
H = \frac{1}{2} \int d^3x \left[\mathbf{E}^a \cdot \mathbf{E}^a + \mathbf{B}^a \cdot \mathbf{B}^a\right],
\tag{6}
$$

is thus

$$
H = \frac{g^2}{a} \left\{ \sum_l \frac{1}{2} E_l^a E_l^a + \frac{2N}{g^4} \sum_p \left[1 - \frac{1}{2N} \mathrm{Tr}\left(U_p + U_p^\dagger\right)\right]\right\}.
\tag{7}
$$

This lattice Hamiltonian for a pure gauge theory was first derived by Kogut and Susskind (1975).

Each link variable U_l, which is a SU(N) matrix, can be regarded as coordinates of a point in the SU(N) group manifold. Since E_l^a is the momentum conjugate to this "position" variable, $E^a E^a$ is just $-\nabla^2$, the Laplace-Beltrami operator on the manifold. The lattice Hamiltonian (7) therefore describes a non-relativistic, quantum many-particle system in the SU(N) group manifold, interacting with each other through a four-body potential $V(U_1, U_2, U_3, U_4) = \frac{2N}{g^4}\left[1 - \frac{1}{2N}\mathrm{Tr}(U_p + U_p^\dagger)\right]$, where U_p is the plaquette formed by the four link variables U_1, U_2, U_3 and U_4. ¿From this perspective, the dynamics takes place entirely in the group manifold and the underlying lattice structure merely serves to identify participants of each four-body interaction.

3 Variational Theory and Glueball Masses

A variational wavefunction for the ground state of (7), which is exact in the strong coupling limit $\xi \equiv 1/Ng^2 \ll 1$, is (Hofsäss and Horsley, 1983)

$$
\Phi_0 = \exp S = \exp\left[AN \frac{1}{2} \sum_p \mathrm{Tr}\left(U_p + U_p^\dagger\right)\right].
\tag{8}
$$

A is the variational parameter. By use of the Jackson-Feenberg (1961) identity in the present context,

$$
\langle -(E_l^a S)(E_l^a S)\rangle = \langle \tfrac{1}{2} E_l^a E_l^a S\rangle,
\tag{9}
$$

where $\langle O \rangle \equiv \langle \Phi_0 \mid O \mid \Phi_0 \rangle / \langle \Phi_0 \mid \Phi_0 \rangle$, which follows from

$$
\langle \Phi_0 \mid E_l^a E_l^a \mid \Phi_0 \rangle = -\langle E_l^a \Phi_0 \mid E_l^a \Phi_0 \rangle,
\tag{10}
$$

the resulting variational ground state energy per plaquette per gluon degree of freedom can be expressed as

$$\varepsilon_0 = \frac{a\langle H \rangle}{N_p(N^2-1)} = \frac{1}{\xi}\left\{\frac{1}{2}AP(A) + \xi^2 \frac{2N^2}{N^2-1}[1 - P(A)]\right\}. \tag{11}$$

$N_p = 3L^3$ is the number of plaquettes (or links) in a L^3 sites lattice. $P(A)$ is the plaquette expectation value defined below. In deriving (11), I have used the fact that $E_l^a U_l = T^a U_l$ and that $T^a T^a = (N^2-1)/2N$. The equivalence (9) also implies that there are two operators,

$$P = \frac{1}{N_p}\frac{1}{2N}\sum_p \text{Tr}(U_p + U_p^\dagger), \tag{12}$$

and

$$\tilde{P} = \frac{AN}{N_p(N^2-1)}\sum_{l,a}\left|E_l^a \frac{1}{2}\sum_p \text{Tr}(U_p + U_p^\dagger)\right|^2, \tag{13}$$

whose expectation values with respect to Φ_0^2, are equal:

$$\langle P \rangle = \langle \tilde{P} \rangle \equiv P(A). \tag{14}$$

In Monte Carlo calculations, it is advantageous to compute $P(A)$ by the use of $\tilde{P}$, which generally has a smaller variance than P.

Minimizing ε_0 with respect to A, yields

$$\xi^2 = \frac{N^2-1}{4N^2}A\left[1 + \frac{P(A)}{AP'(A)}\right], \tag{15}$$

which ostensibly determines $A(\xi)$ implicitly as a function of ξ. However, it is more useful to regard (11) and (15) as parametric equations for determining $\varepsilon_0(\xi)$ in terms of A. Note that the energy optimization has been done analytically, it is only necessary to compute $P(A)$ and $P'(A)$ in order to determine $\varepsilon_0(\xi)$. The derivative $P'(A)$ can be obtained by numerically differentiating $P(A)$, or by computing the variance of P via

$$P'(A) = 2N^2 N_p[\langle P^2 \rangle - \langle P \rangle^2]. \tag{16}$$

In practice, it is easier to do the former (Chin et al., 1986).

For any operator F, one can define a family of sum rules by

$$m_k = \sum_{n\neq 0}(E_n - E_0)^k|\langle 0|F|n\rangle|^2 \tag{17}$$

If F is an operator such that $F|0\rangle$ projects strongly onto the excited state $|1\rangle$, then the corresponding excitation energy is bounded from above by (Bohigas et al., 1979)

$$E_1 - E_0 \leq \frac{m_k}{m_{k-1}}. \tag{18}$$

The most commonly used sum rules are

$$m_0 = \langle 0|F^2|0\rangle - |\langle 0|F|0\rangle|^2 \tag{19}$$

$$m_1 = \frac{1}{2}\langle 0|[F^\dagger, [H, F]]|0\rangle, \tag{20}$$

$$m_3 = \frac{1}{2}\langle 0|[[F^\dagger, H], [H, [H, F]]|0\rangle, \tag{21}$$

and m_{-1}. The latter can be extracted by computing the ground state energy shift. If H is perturbed by αF

$$H' = H - \alpha F \tag{22}$$

then

$$m_{-1} = -\frac{1}{2}\frac{\mathrm{d}^2 E_0'}{\mathrm{d}^2\alpha}\bigg|_{\alpha=0}. \tag{23}$$

The merit of using sum rules is that one can extract excited state properties solely on the basis of ground state expectation values. Thus given a set of m_k's, the excitation energy, or the glueball mass, $M = E_1 - E_0$, can be estimated by minimizing

$$M = \frac{m_1}{m_0}, \tag{24}$$

$$M = \frac{m_0}{m_{-1}}, \tag{25}$$

or

$$M^2 = \frac{m_3}{m_1} \tag{26}$$

with respect to F. A simple F for estimating the O^{++} glueball mass is the single plaquette operator $F = P$, which is exact in the strong-coupling limit. The ground state expectation values required in (24)-(26) can in principle be computed via the Diffusion Monte Carlo algorithm. However, for a quick assessment, these sum rules can be evaluated by replacing the exact ground state by the trial ground state Φ_0. Such an approximation is limited by the quality of Φ_0 and can only be justified *a posteriori*. In this case the m_1 estimate (24) reduces to

$$Ma = g^2\left[\frac{\langle -\frac{1}{2}\sum_l (E_l^a P)(E_l^a P)\rangle}{\langle P^2\rangle - \langle P\rangle^2}\right]. \tag{27}$$

Since the numerator is related to (13) and the denominator is just (16), the above simplifies to (Chin *et al.*, 1986)

$$Ma\xi = \frac{N^2 - 1}{N^2}\frac{P(A)}{AP'(A)}. \tag{28}$$

Thus once $P(A)$ and $P'(A)$ are evaluated to determine the ground state energy, this mass estimate is essentially given without further effort. To obtain the better m_{-1} estimate, let

$$e_0 = \langle H \rangle a\xi = N_p(N^2 - 1)\frac{1}{2}AP(A) + \alpha[1 - P(A)] \qquad (29)$$

where

$$\alpha = 2N^2 N_p \xi^2 = \frac{1}{2}N_p(N^2 - 1)\left[A + \frac{P(A)}{P'(A)}\right], \qquad (30)$$

then m_{-1} according to (23) is given by

$$m_{-1} = -\frac{1}{2}\frac{d^2 e_0}{d^2\alpha} = \frac{1}{2}P'(A)/\frac{d\alpha}{dA}. \qquad (31)$$

The resulting m_{-1} estimate (25) is thus (Chin, 1991)

$$Ma\xi = \frac{N^2 - 1}{N^2}\left[1 - \frac{P(A)P''(A)}{2P'(A)^2}\right]. \qquad (32)$$

Again, (28) or (32) in conjunction with (15) determine $M(\xi)$ parametrically. The required second derivative can also be obtained by numerical differentiation.

The variational ground state (8) can be systematically improved by including more and larger gauge-invariant Wilson loops U_i in the form

$$\Phi_0 = \exp\left[NN_p \sum_i A_i O_i\right], \qquad (33)$$

where

$$O_i = \frac{1}{N_i}\frac{1}{2}\sum \text{Tr}\left[U_i + U_i^\dagger\right]. \qquad (34)$$

N_i are the numbers of such loops in a 3-D lattice. In the next section, I will present results with the addition of three 6-links Wilson loops. Similarly, glueball masses can be improved by parametrizing the excited state projector as

$$F = \sum_i C_i O_i \qquad (35)$$

and minimizing $M = E_1 - E_0$ with respect to coefficients C_i. More details can be found in (Chin, 1991)

4 Exact Ground State Energy, Scalar Glueball Mass and Mass Ratios

The exact ground state energies of U(1), SU(2) and SU(3) Hamiltonian lattice gauge theories in 3-D have been systematically determined by the Diffusion Monte Carlo algorithm (Chin et al., 1984, 1985, 1988b). Since only SU(3) is of physical interest, I will summarize results for this case only.

4.1 Ground State Energy

Fig. 1 shows the SU(3) ground state energy per plaquette per gluon degree of freedom in units of $1/a$. The dotted line is the one-plaquette variational energy (11) and (15). The hollow plotting symbols are exact Diffusion Monte Carlo results (Chin *et al.*, 1988b). As one can see, the variational results are exact in the strong-coupling limit where $\xi \approx 0$ but is substantially higher in the weak-coupling limit. In the physical scaling region, around $\xi \gtrsim 0.35$, where the lattice results are supposed to match onto the continuum limit, the deviations are still noticeable. By improving the variational ground state to (33) with the addition of three 6-links Wilson loops as shown in Fig. 2, the energy can be lowered into agreement with DMC results. These new variational energies (Chin, 1991) are plotted as solid circles. The improvement is substantial, but the work required to optimize and sample this complicated trial function is an order of magnitude greater than that of the single plaquette trial function (8).

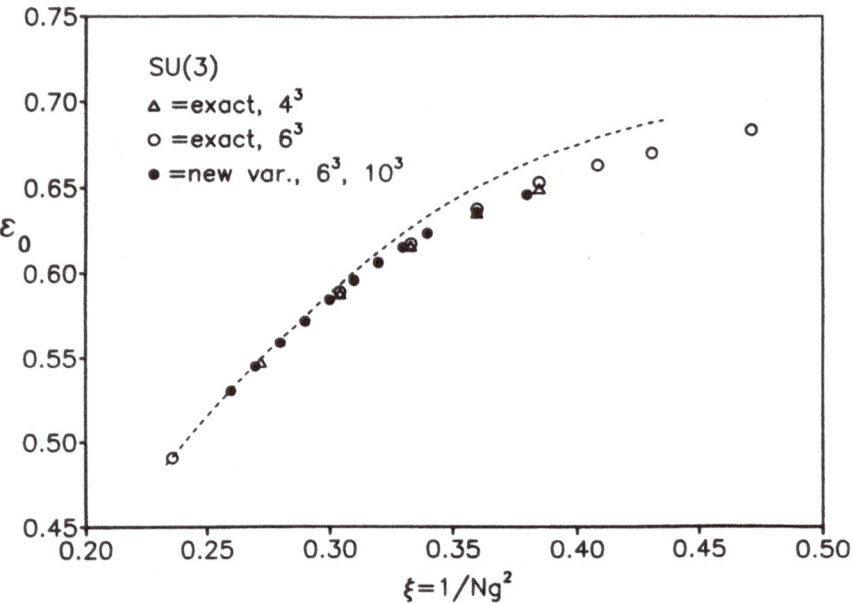

Fig. 1. SU(3) Hamiltonian lattice ground state energy in 3-D. ε_0 is the dimensionless energy per plaquette per gluon defined by $\varepsilon_0 = a\langle H\rangle/[N_p(N^2 - 1)]$.

4.2 The Scalar Glueball Mass

In a lattice gauge theory, in order for any mass to have a fixed value in the continuum limit, g^2 must vary with the lattice spacing a in a well defined manner governed by the β function of theory. When the continuum limit is approached by letting $a \to 0$, asymptotic freedom requires that g^2 must go to zero as well.

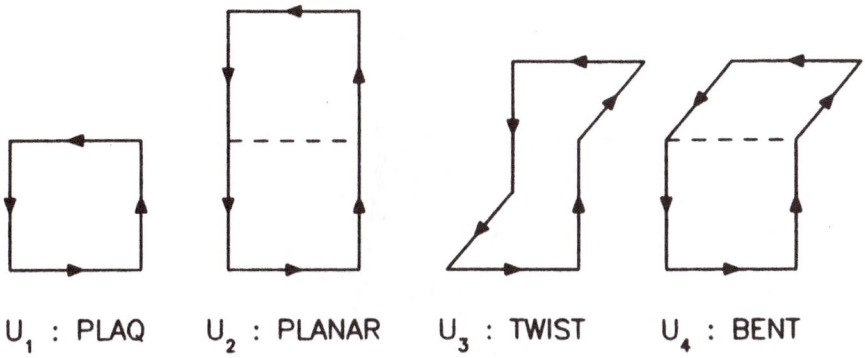

Fig. 2. The 4- and 6-links Wilson loops used in variational calculations.

Thus the continuum limit of a lattic gauge theory is to be sought in the weak-coupling regime. The precise manner in which both a and g^2 must tend to zero can be computed perturbatively in the continuum theory and is given in terms of a lattice mass scale parameter Λ_L

$$\Lambda_L a = \left(\frac{48\pi^2}{11}\xi\right)^{51/121} \exp\left(-\frac{24\pi^2}{11}\xi\right). \tag{36}$$

In the continuum limit, any mass in a SU(N) gauge theory must be proportional to Λ_L in its dependence on the coupling constant g^2. Fig. 3 gives the scalar glueball mass based on the m_1 sum rule (28) for SU(3) to SU(6) (Chin *et al.*, 1986). Lines proportional to Λ_L are shown as two nearly vertical straight lines. For SU(3) and SU(4), there is no indication that the resulting glueball mass exhibits such a proportional, or scaling, behavior. However, scaling begins to emerge for SU(5) and SU(6). This calculation thus demonstrates that a variational Hamiltonian lattice calculation can produce scaling behavior.

The m_{-1} estimate (32) yielded lower glueball masses for SU(4) to SU(6) and suggested scaling even for SU(4). Unfortunately, no improvement was observed for SU(3) (Chin, 1991). The single plaquette variational theory appears to work better at larger values of N. This is in general agreement with $1/N$ expansion studies.

The scalar glueball mass can also be improved with a better excited state projector. Fig. 4 shows the resulting improvement when some or all of the operators in Fig. 2 are used in (35). For comparisons, I have replotted the non-scaling, single plaquette variational results of Fig. 3 as asterisks. The solid squares are

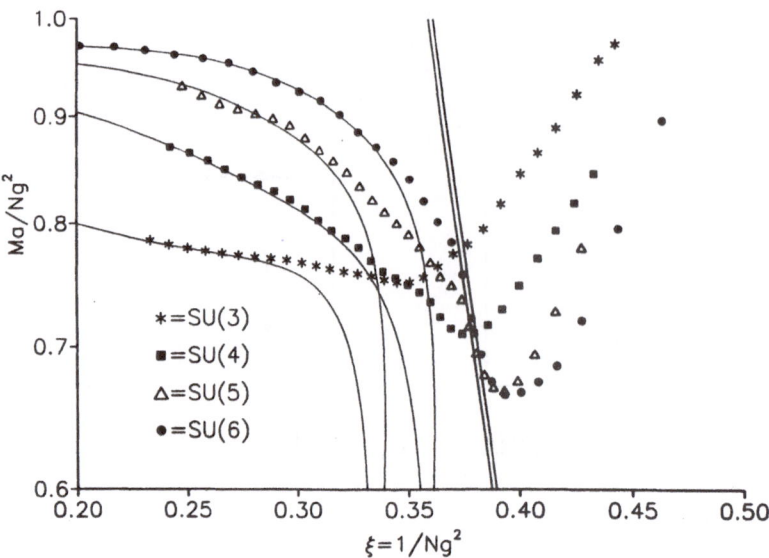

Fig. 3. SU(N) scalar glueball masses as obtained by using the single plaquette m_1 sum rule estimate (28).

results from improving both the trial ground state and the excited state projector with U_1 and U_3 (Long *et al.*, 1988; Robson *et al.*, 1989). The hollow triangles use all four Wilson loops in the excited state projector, but only retain U_1 in the trial ground state. The greatest improvement is obtained when both the ground and the excited state are optimized with all four Wilson loops (Chin, 1991). These are shown as solid and hollow circles for two lattice sizes. The scaling lines drawn give

$$M(0^{++}) = (450 \pm 30)\Lambda_L, \tag{37}$$

which is comparable to Lagrangian lattice results using the same number of Wilson loops.

4.3   The $M(0^{++})/\sqrt{\sigma}$ Ratio

If two masses both scale over the same range of g^2 according to (36), then their ratio is independent of both g^2 and a reflecting the continuum mass ratio. It is possible that over a different range of g^2, depending on the nature of the approximation, both masses may scale with some *non − perturbative* scaling function other than the continuum scaling function (36). In this case their ratio would also be constant over this range of g^2, reproducing the continuum ratio, but each mass individually may not be proportional to Λ_L. Thus mass ratios are more robust in extrapolating to their continuum values. In Fig. 5, the ratios of the scalar glueball masses (that of Fig. 3) to the square root of the string tension $\sqrt{\sigma}$, are plotted (Chin and Karliner, 1987). Values for σ are taken from

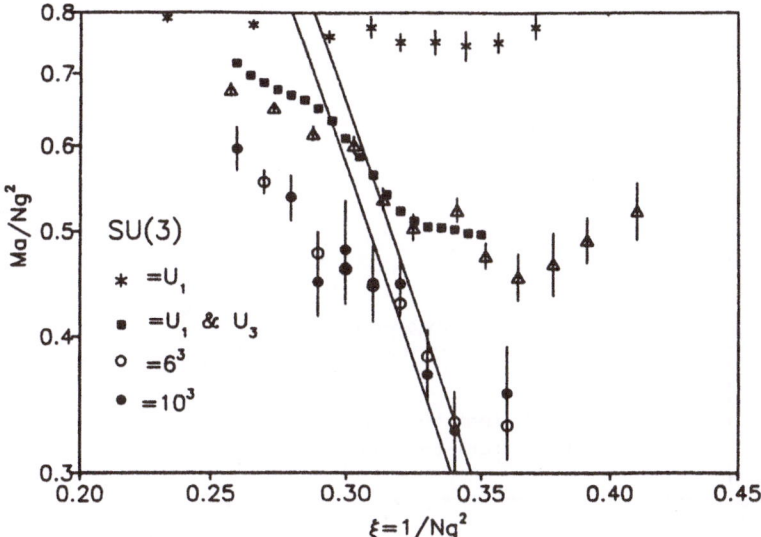

Fig. 4. SU(3) scalar glueball masses with improved ground state wavefunctions and excited state projectors. See text for detail.

the strong coupling expansion of Kogut and Shigemitsu (1980). (They did not give results for the case $N = 4$). The ratios are all approximately constant ≈ 3, over a fair range of ξ, in good agreement with Lagrangian lattice calculations.

4.4 The $M(2^{++})/M(0^{++})$ Ratio

The chief advantage of the Hamiltonian formulation is that it directly addresses the question of the spectrum. Just as one can project out the scalar glueball mass by use of the single plaquette operator

$$F = P \equiv P_x + P_y + P_z, \qquad (38)$$

the tensor glueball mass, in the same approximation, can be projected out by taking (Kogut *et al.*, 1976)

$$F = P_x - P_y, \qquad (39)$$

where P_x, P_y and P_z are lattice sums of plaquettes perpendicular to the x, y and z directions respectively. The resulting mass ratios for SU(3) and SU(6) (Chin *et al.*, 1988a) are shown in Fig. 6. First, one notices that these ratios are always greater than one. Despite some early controversy (Berg *et al.*, 1986), there is no indication that the tensor mass is lower than the scalar mass. Second, although there is no indication of a constant scaling window, there is a slightly flattened, hill-like structure in SU(3). The ratio at that point, ≈ 1.5, also agrees well with other more refined calculations of this ratio (Michael and Teper, 1988). Thus despite the crudeness of the individual projector, the resulting mass ratio is surprisingly reasonable.

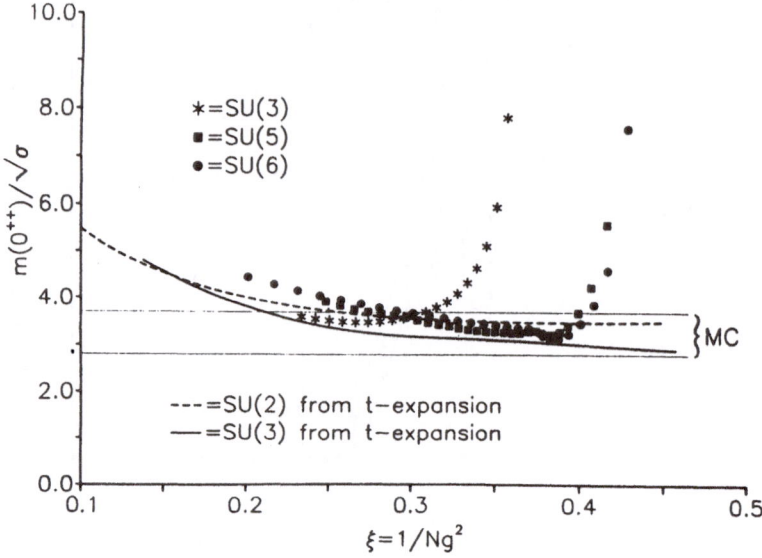

Fig. 5. The $M(0^{++})/\sqrt{\sigma}$ ratio as computed in SU(N) Hamiltonian lattice theories. Plotting symbols denote ratios of the single plaquette glueball mass of Fig. 3 to $\sqrt{\sigma}$. See text for details.

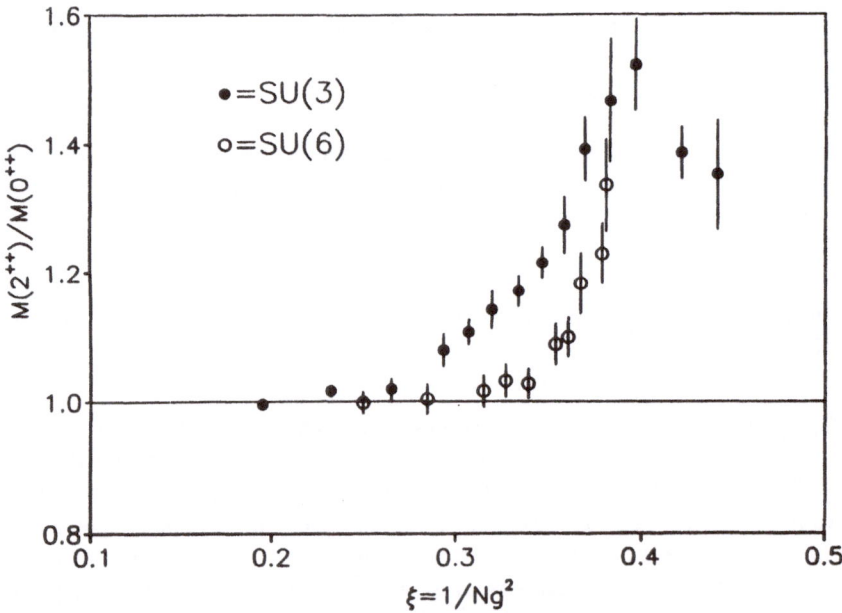

Fig. 6. The tensor-to-scalar glueball mass ratio as calculated in the single plaquette approximation.

5 Latticized Chiral Meson Field Theories

Chiral symmetry is important for understanding the hadronic dynamics of pions and nucleons. Recent interests in heavy-ion collisions have focused attention on the nature of the chiral phase transition, equation of state of a hot pion gas (Bunatian and Wambach, 1994), and the temperature/density dependence of meson masses (Brown and Rho, 1991). The first of these topics has been studied extensively in Lagrangian lattice field theory (Kogut et al., 1982) and chiral perturbation theory (Gerber and Leutwyler, 1989). Since the Hamiltonian approach directly addresses the question of the spectrum, it is an ideal tool for studying pion energetics and pionic collective excitations. In this work, I will report some recent Hamiltonian lattice studies on the ground and collective states of the sigma model.

The Lagrangian density for the non-linear sigma model can be taken to be

$$L = \frac{f_\pi^2}{4} \operatorname{Tr}\left(\partial_\mu U^\dagger \partial_\mu U\right) - m_\pi^2 f_\pi^2 (1 - \tfrac{1}{2}\operatorname{Tr}U), \tag{40}$$

where U is now an *isospin* SU(2) matrix,

$$U = \exp(i\boldsymbol{\tau} \cdot \boldsymbol{\pi}/f_\pi). \tag{41}$$

By expanding in powers of f_π^{-1}, one obtains

$$L = \frac{1}{2}\left(\partial_\mu \boldsymbol{\pi}\right)^2 - \frac{1}{2}m_\pi^2 \boldsymbol{\pi}^2 + \frac{1}{2f_\pi^2}\left(\boldsymbol{\pi} \cdot \partial_\mu \boldsymbol{\pi}\right)^2 - \frac{m_\pi^2}{8f_\pi^2}(\boldsymbol{\pi}^2)^2 + \cdots, \tag{42}$$

which correctly reproduces the low energy scattering of pions to order f_π^{-2}. To latticize (40) over a cubic lattice of spacing a, one simply replaces spatial derivatives by finite differences

$$\partial_i U^\dagger \partial_i U \;\rightarrow\; \frac{1}{a^2}\left[U^\dagger(\mathbf{r}+a\mathbf{i}) - U^\dagger(\mathbf{r})\right]\left[U(\mathbf{r}+a\mathbf{i}) - U(\mathbf{r})\right] \tag{43}$$

$$\rightarrow\; \frac{2}{a^2}\left[1 - U(\mathbf{r})U^\dagger(\mathbf{r}+a\mathbf{i})\right], \tag{44}$$

and replace the corresponding conjugate momentum by the operator E^a. I will first study the case without the symmetry-breaking mass term. Its effect can be incorporated later if necessary. The resulting lattice Hamiltonian is then simply (see also McLerran and Willensen, 1976)

$$H = \frac{1}{a}\frac{4}{(f_\pi)^2}\left\{\sum_{\mathbf{r}} \frac{1}{2}E^a(\mathbf{r})E^a(\mathbf{r}) + \frac{1}{4}(f_\pi a)^4 \sum_{\mathbf{r},i} \frac{1}{2}\operatorname{Tr}\left[1 - U(\mathbf{r})U^\dagger(\mathbf{x}+a\mathbf{i})\right]\right\}. \tag{45}$$

Note that here, both E_i^a and U_i are defined on lattice sites and the potential energy term only couples nearest neighbor pairs of Us.

6   Variational Ground State and Energy

To study the ground state of (45), it is useful to consider the dimensionless Hamiltonian

$$\tilde{H} = \sum_i \frac{1}{2} \mathbf{E}_i \cdot \mathbf{E}_i + \lambda \sum_{<ij>} \frac{1}{2} \mathrm{Tr} \left[1 - U_i U_j^\dagger \right], \tag{46}$$

where i and $\langle i, j \rangle$ denote lattice sites and nearest neighbor pairs. Each SU(2) matrix at each site, $U_i = x_i^0 - i\tau^a \cdot x_i^a$, can be conveniently labeled by a unit 4-D vector $(x_i^0, \mathbf{x}_i)$ such that $(x_i^0)^2 + \mathbf{x}_i^2 = 1$. The effect of $\mathbf{E}_i$ on U_j is just

$$\mathbf{E}_i U_j = \delta_{ij} \frac{\tau}{2} U_i. \tag{47}$$

The potential energy term in (46) corresponds to

$$\frac{1}{2} \mathrm{Tr}\, U_i U_j^\dagger = x_i^0 x_j^0 + \mathbf{x}_i \cdot \mathbf{x}_j = \cos(\theta_{ij}), \tag{48}$$

where θ_{ij} is the angle between two 4-D unit vectors x_i and x_j. The chiral lattice Hamiltonian (46) thus corresponds to a system of quantum rotors in a 3-D lattice.

As in Section 3, the form of the potential energy naturally suggests the following trial ground state wavefunction:

$$\Phi_0 = \exp \left[A \sum_{<ij>} \mathrm{Tr}\, \frac{1}{2} (U_i U_j^\dagger) \right]. \tag{49}$$

The resulting variational energy can be derived precisely as in the lattice gauge case. Instead of the plaquette expectation value, one now needs the cosine expectation value:

$$C(A) = \frac{1}{N_l} \left\langle \sum_{<ij>} \mathrm{Tr}\, \frac{1}{2} (U_i U_j^\dagger) \right\rangle, \tag{50}$$

where $N_l = 3L^3$ is the number of links in a cubic lattice of L^3 sites. The energy per link is simply

$$\varepsilon_0(\lambda) = \frac{\langle \tilde{H} \rangle}{N_l} = \lambda + \left[\frac{3}{16} A - \lambda \right] C(A). \tag{51}$$

Minimizing this with respect to A again determines λ and $\varepsilon_0(\lambda)$ parametrically in terms of A,

$$\lambda = \frac{3}{16} \left[A + \frac{C(A)}{C'(A)} \right]. \tag{52}$$

The cosine expectation value and its derivative are shown in Fig. 7 for an 8^3 lattice. The resulting ground state energy per link is shown in Fig. 8. Also shown are $C(\lambda)$ and $1 - \frac{d\varepsilon_0}{d\lambda}$, which by the Feynman-Hellman theorem, must be equal. Their equality is used to check the accuracy of various numerical derivatives.

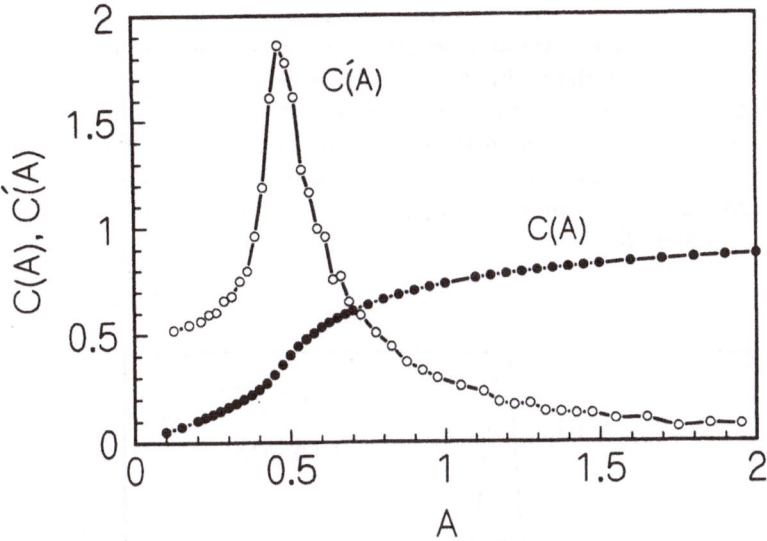

Fig. 7. The cosine expectation value $C(A)$ and its derivative $C'(A)$ as functions of the variational parameter A.

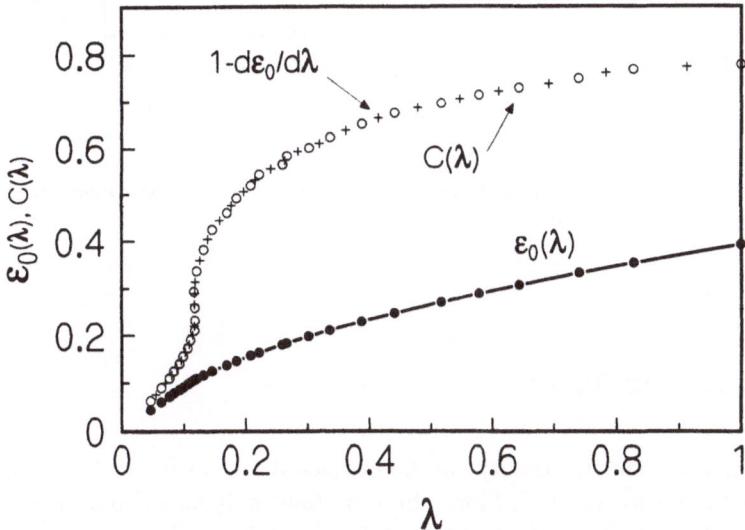

Fig. 8. The variational ground state energy per link ε_0 as a function of the dimensionless chiral coupling constant $\lambda = \frac{1}{4}(f_\pi a)^4$. Also shown are the cosine expectation value $C(\lambda)$ and the numerical derivative $1 - d\varepsilon_0/d\lambda$.

Fig. 9 gives an expanded view of the ground state energy and its derivatives near $\lambda \approx 0.12$. For $\lambda < 0.12$, the cosine expectation is close to zero, implying that all 4-D vectors are independently and randomly oriented. This is the symmetric chiral phase. For $\lambda > 0.12$, the cosine expectation is close to one, suggesting that all the 4-D rotors are closely aligned and pointing in the same direction. This is the spontaneously broken symmetry phase. The analogy with ferromagnetic phase transitions is obvious. The second order nature of the transition is also clear; the second derivative of the energy diverges. (For a finite lattice, this divergence is rounded off.)

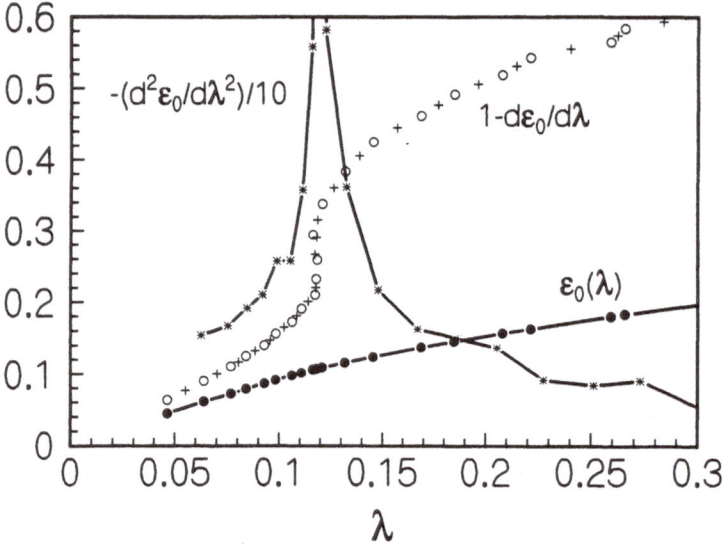

Fig. 9. The ground state energy and its first and second derivatives near the chiral symmetry phase transition point.

7   The ω Meson Mass

Collective excitations in this theory can be projected out, as in the lattice gauge case, with suitable choices of F. Since the pion field π is an isospin vector, the following two choices will project out a vector-isoscalar and a vector-isovector collective mode respectively:

$$F = \begin{cases} \boldsymbol{\pi} \cdot \nabla_i \boldsymbol{\pi}, & \text{isoscalar} \\ \boldsymbol{\pi} \times \nabla_i \boldsymbol{\pi}. & \text{isovector} \end{cases}$$

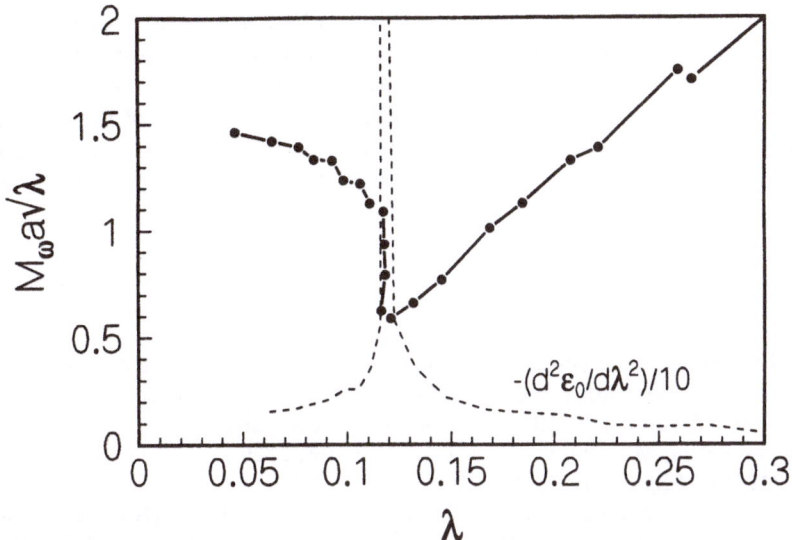

Fig. 10. The mass gap of the ω collective mode as a function of the dimensionless chiral coupling constant λ. The dashed line gives the second derivative of the ground state energy.

I will refer to these two as the ω and the ρ meson modes. Since replacing derivatives by finite differences leads to nearest neighbor coupling,

$$\boldsymbol{\pi} \cdot \nabla_i \boldsymbol{\pi} \;\to\; \boldsymbol{\pi}(\mathbf{r}) \cdot \boldsymbol{\pi}(\mathbf{r} + a\mathbf{i}) + \cdots,$$
$$\to \text{Tr}\big[\tau^a U(\mathbf{r})\big]\,\text{Tr}\big[U(\mathbf{r} + a\mathbf{i})\tau^a\big],$$
$$\to \text{Tr}\big[\tau^a U(\mathbf{r})U(\mathbf{r} + a\mathbf{i})\tau^a\big] + \cdots,$$
$$\to \text{Tr}\big[U(\mathbf{r})U(\mathbf{r} + a\mathbf{i})\big],$$

we can therefore take

$$F_\omega = \sum_{<ij>} \tfrac{1}{2}\text{Tr}\big[U_i U_j^\dagger\big], \tag{53}$$

and

$$F_\rho = \sum_{<ij>} \tfrac{1}{2}\text{Tr}\big[\tau^a U_i U_j^\dagger\big]. \tag{54}$$

The ω meson mass is therefore exactly analogous to the single plaquette scalar glueball mass. The corresponding m_1 estimate gives

$$M_\omega a\sqrt{\lambda} = \frac{3}{2}\frac{C(A)}{AC'(A)}. \tag{55}$$

This is plotted in Fig. 10. Since $M_\omega \propto 1/C'(\lambda) \propto 1/(d^2\varepsilon_0/d\lambda^2)$, it is only because of the finite lattice size that M_ω is prevented from plunging to zero. This decrease in meson masses as one approaches the chiral phase transition point has been suggested previously by Brown and Rho (1991). Their argument was based on the mass scaling behavior of some effective chiral lagrangians. It is interesting to see that this effect emerges naturally in a fundamental calculation. This softening of the ω collective mode is also strikingly similar to the softening of the magnon mass gap in spin systems (Ristig and Kim, 1996 and reference therein; also their contribution to this workshop). The temperature dependence of both the ω and the ρ mass will be discussed in a future publication.

References

Berg, B. A., Billoire, A., and Vohwinkel, C. (1986): Phys. Rev. Lett. **57**, 400 .

Bohigas, O., Lane, A., and Martorell, J. (1979): Phys. Rept. **51**, 267 .

Bunatian, G. and Wambach, J. (1994): Phys. Lett. **B336**, 290 .

Brown, G. E. and Rho, M. (1991): Phys. Rev. Lett. **66**, 2720 .

Chin, S. A. (1978): Phys. Lett. **B78**, 552 .

Chin, S. A., Negele, J. W. and Koonin, S. E. (1984): Ann. of Phys. (NY) **157**, 140 .

Chin, S. A., van Roosmalen, O. S. , Umland, E. A., and Koonin, S. E. (1985): Phys. Rev. **D31**, 3201 .

Chin, S. A. (1986): *Cluster Model and Other Topics*, International Review of Nuclear Physics, Volume 4, edited by T. Kuo, (World Scientific, Singapore) 395-510.

Chin, S. A., Long, C. and Robson, D. (1986): Phys. Rev. Lett. **57**, 2779.

Chin, S. A. and Karliner, M. (1987): Phys. Rev. Lett. **58**, 1803.

Chin, S. A., Long, C. and Robson, D. (1988a): Phys. Rev. Lett. **60**, 1467.

Chin, S. A., Long, C. and Robson, D. (1988b): Phys. Rev. **D37**, 3001.

Chin, S. A. (1991): *Progress in Nuclear Physics*, edited by W-Y. P. Hwang *et al* (Elsevier, New York) 47–63.

Gerber, P. and Leutwyler, H. (1989): Nucl. Phys. **B321**, 387 .

Hofsäss, T. and Horsley, R. (1983): Phys. Lett. **123B**, 65 .

Jackson, H, and Feenberg, E. (1961): Ann. of Phys. (NY) **15**, 266 .

Kogut, J., and Susskind, L. (1975): Phys. Rev. **D11**, 395 .

Kogut, J., Sinclair, D. K. and Susskind, L. (1976): Nucl. Phys. **B114**, 199 .

Kogut, J., and Shigemitsu, J. (1980): Phys. Rev. Lett. **45**, 410 .

Kogut, J., Snow, M. and Stone, M. (1982): Nucl. Phys. **B200[FS4]**, 211 .

Long, C., Robson, D. and Chin, S. A. (1988): Phys. Rev. **D37**, 3006.

McLerran, L. D., and Willemsen, J. F. (1976): Phys. Lett. **65B**, 351 .

Michael, C. and Teper, M. (1988): Phys. Lett. **B206**, 299 .

Ristig, M. L., and Kim, J. W. (1995): Phys. Rev. **B53**, 6665 .

Robson, D., Long, C. and Chin, S. A. (1989): "Hamiltonian Lattice Gauge Calculation for SU(N) Gauge Group", *Relativistic Nuclear Many-Body Physics*, (World Scientific, Singapore) 31-46.

Yang, C. N. (1974): Phys. Rev. Lett. **33**, 445 .

Stochastic Projection of the Ground State of Strongly Correlated Electrons

Efstratios Manousakis

Department of Physics and
Center for Materials Research and Technology
Florida State University, Tallahassee, Florida, 32306, USA

Abstract. An overview is presented of Green's function Monte Carlo methods which have been applied to the two-dimensional $t - J$ model, a simple quantum lattice-gas model which captures strong electron correlations. Exact diagonalization methods are restricted to very small size systems which do not allow extrapolation to the thermodynamic limit. On the other hand, world-line Monte Carlo simulation methods suffer from the fact that the antisymmetric nature of the amplitudes for fermions makes the simulation error grow exponentially with system size. Thus, in straightforward applications of simulation techniques or in exact diagonalization techniques the required computer time grows exponentially with system size. We have developed projection methods similar to the Green's function Monte Carlo technique which can be used to extract the many-body ground state where the above mentioned problems can be kept under control. This technique reproduces all known exact diagonalization results on smaller size systems and it can be applied on significantly larger size lattices. Using this method, the motion of a hole, the hole-binding and the phase diagram of the $t - J$ model have been studied. Among other important results, it has been found that there is phase separation in the physical region of the model. The relevance of the phase diagram of the model to the physics of the cuprate superconductors and the role of additional terms, which need to be added to the model in order to describe the environment experienced by the electrons in the real materials, are briefly discussed.

1 Introduction

Collective behavior can be understood using models idealizing the nature of the environment and the mutual interaction of the interacting constituents. Even though one might have attempted to eliminate all unnecessary complexity in the model itself, in most cases providing an exact solution to the model is of overwhelming difficulty. For example, the antiferromagnetic Heisenberg model

$$H = J \sum_{\langle ij \rangle} \mathbf{S}_i \cdot \mathbf{S}_j, \tag{1}$$

where $\mathbf{S}_i$ are spin-1/2 operators and the sum is over all nearest neighbor pairs on a square lattice, was thought to describe the undoped parent compounds of the superconducting copper oxides. Despite its simplicity this model has no exact solution for an infinite square lattice. However, because one deals with only spin degrees of freedom, numerical solution to the model with satisfactory accuracy has become possible. The emerging physical picture using the Hamiltonian (1) and its extensions obtained by adding small correction terms to it, is in good agreement with the experimental results of the undoped cuprous oxides (Manousakis 1991).

Next one needs to examine what happens when holes are introduced in a minimal way in the Heisenberg antiferromagnet. The so-called t-J model is one of the simplest abstractions, perhaps the simplest one, to describe the environment experienced by the holes (electrons) in the limit of strong on-site Coulomb repulsion. The t-J Hamiltonian, on a square lattice, is written in the subspace with no doubly occupied sites as

$$H = -t \sum_{\langle ij \rangle \sigma} (c_{i\sigma}^{\dagger} c_{j\sigma} + \text{h.c.}) + J \sum_{\langle ij \rangle} (\mathbf{S}_i \cdot \mathbf{S}_j - \frac{n_i n_j}{4}), \tag{2}$$

where $c_{i\sigma}^{\dagger}$ creates an electron of spin σ on site i, and $n_i = \sum_{\sigma} c_{i\sigma}^{\dagger} c_{i\sigma}$. One can hope, by adding correction terms to such a model, some day to reach understanding of the climate in the doped materials. It is clear, however, that to make any progress one needs to develop the capacity of understanding the simple first.

This paper is a review of Green's function Monte Carlo methods used to study correlated electrons and in particular the t-J model both for its simplicity and for its possible relevance to the physics of the copper-oxide superconductors.

Even though numerous exact diagonalization studies (Dagotto 1994) have been performed on small size systems, the situation about the phase diagram of the model is very unclear. Because of the ambiguity in extracting information relevant for the thermodynamic limit of the model from small size systems, the confusion in our community grows with the growing number of such studies. In exact diagonalization studies the dimensionality of the Hilbert space grows exponentially with system size and this limits the diagonalization to lattices of size $L \times L$ with $L \sim 4 - 5$. With periodic boundary conditions, the longest possible distance along a side of the square is only $L/2$ because any correlation function calculated at relative separation $\mathbf{r} = (x, y)$ should be identical to its value at relative vector $\mathbf{r}' = (L - x, y)$. Thus, such a small size lattice has very little resemblance to the $L \to \infty$ lattice.

The Monte Carlo method which can solve large-size classical, quantum spin or boson systems could be considered as an alternative approach. However, in the case of fermions, such stochastic methods are seriously hindered by the so-called sign problem. In a stochastic method one needs to interpret the amplitude for a given configuration as the probability for a walker to be in this configuration as it moves in the many-variable configuration space. For any arbitrary configuration in a fermion system, however, there are configurations very close to it, such as the one that corresponds to itself where a pair is exchanged, with almost the same size amplitude but opposite sign. This leads to cancellation of quantities of similar magnitude each one fluctuating, thus, the cancellation leaves behind a much smaller quantity which fluctuates sharply around zero. This makes the error in the simulation grow exponentially with the system size L^2 which is as catastrophic as the exponential growth with L^2 of the dimensionality of the Hilbert space in exact diagonalizations.

Our approach is a Monte Carlo method, the so-called Green's function Monte Carlo, which also suffers by the above mentioned sign problem. However, this approach allows projection of the exact ground state starting from an initial trial state Ψ_T by repeatedly acting with a projection operator $\hat{P}$. The larger the number of times that $\hat{P}$ is applied to Ψ_T the larger the overlap of the state produced to the true ground state is. Namely, $\hat{P}^n \Psi_T$, is increasingly closer to the true ground state as n increases. On the other hand, the error bars due to the stochastic evaluation of the energy, grow exponentially with n. If, however, the initial state is close to the exact, then, we might achieve convergence to the true ground state for a small enough n so that the errors are not very large. Another factor in the reduction of the error is achieved by using carefully chosen functions to guide the random walk. Thus, this approach is not a straightforward numerical method, the more analytical understanding one has about the system, the better initial and guidance functions one can construct to speed up convergence. Namely, one does not require the stochastic process to give us the total answer; one inputs the known part of the physics and requires the random walkers to simulate only the unknown part and this speeds up the convergence with n. Thus, a necessary first step of this method is a variational approach where accurate wave functions are first optimized to be used as initial and guiding functions to project the true ground state.

This approach is not a universal numerical algorithm that can be developed once and then applied to all systems. It is an approach which requires insight into the physical problem and one is not leaving the entire problem up to the luck of a stochastic numerical approach. It is an approach where one biases the movement of the walker and in the remaining freedom in its movement we seek to understand the part of the physics that is not already known. The entire approach is ergodic and exact within statistical errors provided that we can find convergence with respect to n. When there is convergence we know that we have converged to the lowest possible state of the Hamiltonian with non-zero overlap with the initial state. Thus, different regions of the phase diagram, especially when they correspond to different thermodynamic phases, require different trial states and, thus, different numerical treatment.

In this paper, we examine the phase diagram of the $t-J$ model for almost all interaction strengths and almost all electron densities. The problem is divided into three regions, the region of a few holes, the region of a few electrons and the region of finite density of electrons and holes. In this paper, the results of Green's function Monte Carlo studies of the phase diagram of the $t-J$ model are discussed and their possible relevance for the cuprates is considered.

2 Exact diagonalizations

Let us consider a lattice of N sites and N_u (N_d) total number of electrons with up-spin (down-spin) in a specified direction. Thus, $N_e = N_u + N_d$ is the total number of electrons and $N_h = N_s - N_e$ the total number of holes, and we are interested for electron density $n_e = N_e/N$ less than or at most unity. A straightforward basis of the Hilbert space with well-defined number of electrons $N_u + N_d$ and well-defined magnetization $M = N_u - N_d$ is spanned by

$$|c\rangle = |\mathbf{r}_1, \mathbf{r}_2, ..., \mathbf{r}_{N_u}, \mathbf{R}_1, \mathbf{R}_2, ..., \mathbf{R}_{N_h}\rangle, \tag{3}$$

where we have chosen to use the positions of the up-spin electrons $\mathbf{r}_i$ and the positions of the holes $\mathbf{R}_i$ to specify the configurations. Another equivalent choice is to use the positions of the spin-down electrons to specify the basis.

The Hamiltonian (2) for systems with finite number of spins can be expressed in a basis such as (3) and then diagonalized exactly to obtain all the eigenvalues and eigenstates. The applicability of this approach is limited to small-size systems since the dimensionality of the Hilbert space increases exponentially with the size of the lattice, namely for a lattice with N sites the total number of possible states is 2^N. Considering the global symmetries of H we can restrict ourselves inside invariant subspaces with reduced dimensionality since these subspaces are not coupled by the Hamiltonian. We can work in subspaces characterized by definite eigenvalue of $\hat{S}_{tot}^z$ and S_{tot}^2. In practice the $[H, S_{tot}^z]$ symmetry which leads to reduction in an invariant subspace of well-defined S_{tot}^z and the translational symmetry which leads to a subspace with well-defined momentum can be easily implemented. If the Hamiltonian is a small-size matrix inside the invariant subspaces, it can be diagonalized using standard diagonalization routines, otherwise iterative schemes known as Lanczos algorithms are generally used.

Exact diagonalization studies of the model (2) are limited to sizes $L \times L$ where L is close to 4 or 5. These system sizes are too small, given the fact that with periodic boundary conditions used the longest possible distance along a side of the square is $L/2$. Thus, the correlations at, say, the lattice point (x, y) and $(L - x, y)$ are identical. For an extrapolation to an infinite size lattice one needs a formula for the L dependence of the quantity of interest. Most importantly one needs results obtained for several significantly different size L in order to make sure that such a formula is valid and to extrapolate with confidence in the $L \to \infty$ limit. Since a lattice of size 2×2 has little relation to the $L \to \infty$ case, there is no available range of L's to make the extrapolation possible.

For us, the results of exact diagonalization are very useful because the Monte Carlo method that we are going to develop has to reproduce the known exact results for the small size lattices.

3 Projection methods

The projection method was first used by Metropolis and Ulam (see, e.g., Negele 1988), in which the ground state is projected out using a trial state. This method can be improved significantly by guiding the random walk using "importance sampling" ideas used by Kalos (Kalos et al. 1962, Kalos et al. 1966, Kalos et al. 1981) which shall be explained in the next section. In its first simplest formulation, the method uses a trial state $|\Psi_T>$ which has non-zero overlap with the true ground state $|\Phi_0>$; then one repeatedly applies a convenient projection operator $\hat{P}(\hat{H})$ which projects out the lowest energy eigenstate of the Hamiltonian $\hat{H}$ having non-zero overlap with the trial state. For the case of our lattice models where the spectrum of $\hat{H}$ has an upper bound, a convenient choice is, $\hat{P}(\hat{H}) = E - \hat{H}$, where E is an upper bound to the spectrum of $\hat{H}$. It is straightforward to show that the state

$$|\Psi_n\rangle = \left(\hat{P}(\hat{H})\right)^n |\Psi_T\rangle, \tag{4}$$

or equivalently

$$|\Psi_n\rangle = \hat{P}(\hat{H})|\Psi_{n-1}\rangle, \tag{5}$$

where $|\Psi_0\rangle = |\Psi_T\rangle$, provides a series of increasingly accurate approximations to the lowest eigenstate of $\hat{H}$ having non-zero overlap with $|\Psi_T\rangle$. This can be shown by pretending that we know the set of eigenstates $\{|\Phi_i\rangle\}$ of $\hat{H}$ which form a complete set and expanding the trial state in this basis:

$$|\Psi_T\rangle = c_0|\Phi_0\rangle + c_1|\Phi_1\rangle + ... \tag{6}$$

Then, using (4) and (6) we find:

$$|\Psi_n\rangle \propto |\Phi_0\rangle + \frac{c_1}{c_0}\left(\frac{E - E_1}{E - E_0}\right)^n |\Phi_1\rangle + ..., \tag{7}$$

where E_i are the energy eigenvalues corresponding to the eigenstates $|\Phi_i\rangle$. Since $E > E_i$, the ratio

$$\frac{E - E_i}{E - E_0} < 1, \tag{8}$$

and thus for large n only $|\Phi_0\rangle$ has a non-negligible weight in $|\Psi_i\rangle$.

There are several choices for such a projection operator $\hat{P}$ depending on the investigator and the physical problem. For example, the resolvent operator $\hat{P}(\hat{H}) = \frac{E-z}{\hat{H}-z}$, with $z < E_0$, has been used in certain of the Green's function Monte Carlo (GFMC) studies from which the name originates. For the case of a Hamiltonian with spectrum bounded from above and below such as our lattice models (on a finite lattice), the choice $\hat{P} = \hat{H} - E$ or $E - \hat{H}$ can be made and the matrix elements of $\hat{P}$ can be calculated exactly.

Now we wish to solve Eq. (5) stochastically. We express the matrix elements using the basis $|c\rangle$ defined by Eq. (3). The equation (5) takes the following form

$$\Psi_n(c) = \sum_{c'}\langle c|\hat{P}(\hat{H})|c'\rangle\Psi_{n-1}(c'). \tag{9}$$

Therefore, we can define a stochastic process which produces the distribution Ψ_n starting from Ψ_{n-1}. We begin with a population of N_c walkers. Each walker moves in the configuration space and its initial position is a configuration drawn from the distribution $\Psi_0(c) = \Psi_T(c)$. In the nth iteration, the nth generation of random walkers should be distributed according to $\Psi_n(\mathbf{R})$. The random walker moves from c to c' with a transition probability proportional to $T(c \rightarrow c') \propto \langle c|\hat{P}|c' \rangle$. This approach would work, provided that all the matrix elements of $\hat{P}$ in the chosen basis are positive definite in order to be interpreted as transition probabilities in the stochastic approach. This can always be done for a system of bosons. It can also be done for the case of the Heisenberg antiferromagnet on the square lattice by a suitable transformation of the basis.

In the case of fermions, however, because of the antisymmetric nature of the Hilbert space, a positive-definite transition amplitude for all possible states does not exist. In such case what one can do is to separate the sign from the amplitude and, thus, every walker in the stochastic process carries a flag with a sign (positive and negative walkers) as it moves in the configuration space. In a particular transition where the matrix element of $\hat{P}$ is negative the sign carried by the walker changes. When an average of an operator is calculated each walker contributes with a sign and, thus, a cancellation occurs between the contributions from positive and negative walkers. Because of the nature of the problem, configurations very close to each other (which differ by a pair-permutation) contribute with the same amplitude but with opposite sign. This makes the statistical fluctuations much larger, on the average, than those in a similar calculation for bosons. In fact the longer one proceeds in the power of n to obtain more and more accurate approximations to the true ground state (see Eq. (5)) the larger the error because of the higher power of the hamiltonian involved in the expectation value of an operator. In fact the statistical error grows exponentially with n.

In principle, one can calculate numerically exact ground state expectation values of an operator $\hat{A}$ as follows:

$$\langle \Phi_0|\hat{A}|\Phi_0 \rangle = \lim_{n \to \infty} \frac{\langle \Psi_n|\hat{A}|\Psi_n \rangle}{\langle \Psi_n|\Psi_n \rangle}. \tag{10}$$

In the particular case of the energy expectation value, it is easier to use the expression:

$$E_0 = \frac{\langle \Psi_T|\hat{H}|\Phi_0 \rangle}{\langle \Psi_T|\Phi_0 \rangle} = \lim_{n \to \infty} \frac{\langle \Psi_T|\hat{H}|\Psi_n \rangle}{\langle \Psi_T|\Psi_n \rangle} = \lim_{n \to \infty} \frac{\sum_c \langle c|\Psi_n \rangle \langle \Psi_T|\hat{H}|c \rangle}{\sum_c \langle c|\Psi_n \rangle \langle \Psi_T|c \rangle} \tag{11}$$

$$= \lim_{n \to \infty} \frac{\sum_{c \in \Psi_n} \langle \Psi_T|\hat{H}|c \rangle}{\sum_{c \in \Psi_n} \langle \Psi_T|c \rangle}. \tag{12}$$

The last part of this equation means that one needs to calculate the numerator and denominator separately. The sum is over all those configurations generated in the process of the random walk described previously and which are drawn from the distribution $\Psi_n(c) = \langle c|\Psi_n \rangle$. Given an analytic form of the initial trial state Ψ_T the calculation of the matrix elements $\langle \Psi_T|\hat{H}|c \rangle$ and $\langle \Psi_T|c \rangle$ for a given configuration c is straightforward. It is advantageous to use the expression (12) instead of the expression (10) because the use of Eq. (10) requires forward walking (see, e.g., Negele 1988).

4 Guiding the random walkers

The projection operator moves each particle to any neighbouring site with the same probability. The procedure of obtaining configurations distributed according to the ground state wave function becomes very efficient when the walkers are guided in their moves by a reasonable guiding wave function ψ_G. This method is called "importance sampling" or more commonly Green's function Monte Carlo (GFMC) method. This was an important step forward in treating the ground state of quantum many-body systems using a trial wave function and it was originally developed by Kalos. It has been successfully applied to several continuum systems including liquid helium, helium droplets, molecular systems, liquid and solid hydrogen, nuclear systems and in lattice models including lattice gauge theories.

The guiding action of the wave function reduces the statistical fluctuations and this allows simulation of much larger systems. To see how this works we multiply both sides of Eq. (9) by the guiding function $\Psi_G(c)$ and we rewrite it as

$$\phi_n(c) = \sum_{c'} P_M(c, c') \phi_{n-1}(c'),$$ (13)

where $\phi_n(c) = \Psi_G(c)\Psi_n(c)$ and $P_M(c, c') = \Psi_G(c)P(c, c')/\Psi_G(c')$. Hence, the random walkers are biased and, because of the ratio $\Psi_G(c)/\Psi_G(c')$, certain regions of the configuration space are sampled more often. The price we have to pay is that the distribution of the walkers in the limit $n \to \infty$, is $\phi(c) = \Phi_0(c)\Psi_G(c)$. The ground state energy can be easily calculated by rewriting equation (12) as

$$E_0 = \frac{\sum_c \Phi_0(c)\Psi_G(c) \frac{(H\Psi_T(c))^*}{\Psi_G(c)}}{\sum_c \Phi_0(c)\Psi_G(c) \frac{\Psi_T(c)^*}{\Psi_G(c)}} = \frac{\sum_{c\in\phi(c)} \frac{(H\Psi_T(c))^*}{\Psi_G(c)}}{\sum_{c\in\phi(c)} \frac{\Psi_T(c)^*}{\Psi_G(c)}}.$$ (14)

Therefore, the expectation value (12) can be calculated as the ratio of the averages of $\frac{(H\Psi_T(c))^*}{\Psi_G(c)}$ and of $\frac{\Psi_T(c)^*}{\Psi_G(c)}$ over configurations c generated by (13) for large enough n.

In the "importance sampling" technique, it is very important to have accurate variational wave functions. In the systems where GFMC has been applied with success there was a prior or parallel search for accurate variational wave functions. In addition, having a trial wave function and using the GFMC approach we can determine the accuracy of this wave function. Furthermore, an accurate wave function gives us a useful insight into the dominant physical processes which are important in the determination of the properties of the system.

5 Single hole in an infinite lattice

The no-hole case, namely the ground state of the spin-1/2 Heisenberg antiferromagnet, as has already been discussed, presents no sign problem. One can define a suitable basis where all matrix elements of $\hat{P}$ are positive. Very accurate GFMC calculations for the no hole case have been performed (Carlson 1989, Trivedi and Ceperley 1989, Runge 1992) and they confirm the idea that the ground state possesses long range order (Manousakis 1991).

First we shall discuss the variational ansatz Boninsegni and Manousakis (BM) (Boninsegni and Manousakis 1992a, Boninsegni and Manousakis 1992b) proposed for a single hole in the 2D $t - J$ model, which has been also used as starting trial state for their GFMC calculation. The ansatz for a single-hole can be expressed as follows:

$$|\Psi_T(\mathbf{k})> = \sum_c (-1)^{L(c)} \hat{F}(\mathbf{k}) \; exp\left(-\frac{1}{2}\sum_{i<j} u_{ij}\hat{s}_i^z\hat{s}_j^z\right) \hat{c}_{\mathbf{k}s}|c> , \qquad (15)$$

where $|c\rangle = |\{s_i^z\}\rangle$ is a spin configuration; the sum is restricted to spin configurations with zero z component of the total spin. $L(c)$ is equal to the number of "down" spins in one of the two sublattices; $\hat{c}_{\mathbf{k},s} = 1/\sqrt{N}\sum_{\mathbf{r}} e^{-i\mathbf{k}\cdot\mathbf{r}} \hat{c}_{\mathbf{r},s}$, where the sum runs over all lattice sites. The operator $exp\left(-\frac{1}{2}\sum_{i<j} u_{ij}\hat{s}_i^z\hat{s}_j^z\right)$ is a spin-spin correlation operator and the function u_{ij} depends on the distance between the two sites i and j. The operator $\hat{F}(\mathbf{k})$ is given as

$$\hat{F}(\mathbf{k}) = 1 + \sum_{\mathbf{a}} f_{\mathbf{a}}(\mathbf{k})\hat{\mathcal{P}}_{\mathbf{a}} + \sum_{\mathbf{a},\mathbf{a}'} f_{\mathbf{aa}'}(\mathbf{k})\hat{\mathcal{P}}_{\mathbf{a}'}\hat{\mathcal{P}}_{\mathbf{a}}. \qquad (16)$$

Here $\hat{\mathcal{P}}_{\mathbf{a}} = \sum_{\mathbf{r},s}\hat{c}_{\mathbf{r},s}^{\dagger}\hat{c}_{\mathbf{r}+\mathbf{a},s}$ and $\mathbf{a} = \pm\hat{x},\pm\hat{y}$, namely, it connects two nearest-neighboring sites. Here $\hat{F}(\mathbf{k})$ is a spin-hole correlation operator and $f_{\mathbf{a}}(\mathbf{k})$, $f_{\mathbf{aa}'}(\mathbf{k})$ are variational parameters.

First of all, this trial state takes into account the background spin-spin correlations which are incorporated by the Marshall-Jastrow factor

$$f_{ij} = exp\left(-\frac{1}{2}u_{ij}\hat{s}_i^z\hat{s}_j^z\right), \qquad (17)$$

for all different pairs of spins. In addition, the wavefunction includes spin-hole and spin-hole-spin correlations. The latter correlations describe the "strings" of spins dispaced along the hole path. This wavefunction provides a generalization of the Brinkman-Rice approach to take into account the background spin fluctuations. The variational calculation performed with the optimized wave function (Boninsegni and Manousakis 1992b) gives a good approximation for both the energy of the hole and its spectral weight. The spectral weight is the overlap of the quasihole state (15) to a state describing a non-adiabatically created hole in an antiferromagnet, i.e. to a hole created suddenly before neither the hole nor the spins find the time to relax.

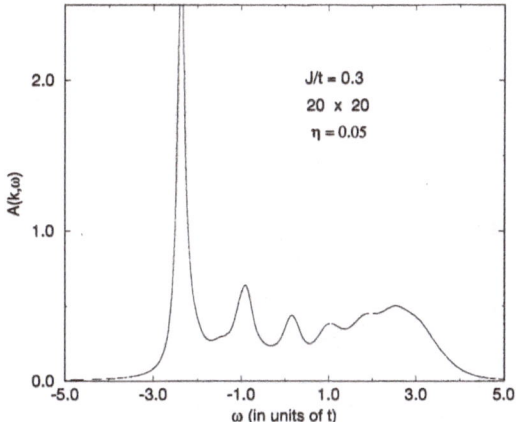

Fig. 1. The hole spectral function $A(\mathbf{k} = (\pi/2, \pi/2), \omega)$ calculated for various hole momenta using the loop expansion (Liu and Manousakis 1992).

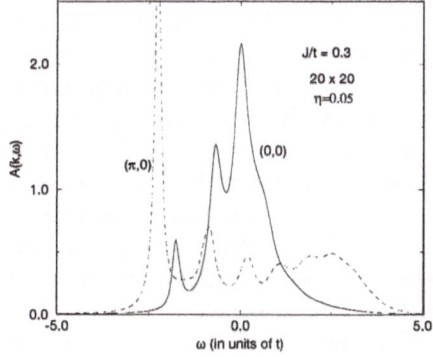

Fig. 2. The hole spectral function for $\mathbf{k} = (0,0)$ and $(\pi, 0)$ calculated using the loop expansion (Liu and Manousakis 1992).

The variational and the GFMC calculation of BM confirmed the results of Liu and Manousakis (Liu and Manousakis 1992) for the single hole energy obtained by a self-consistent loop-expansion (up to two loops). A typical spectral function found by the latter calculation is shown in Fig. 1 and Fig. 2. When $t \gg J$, the typical time-scale for the hole to move from one site to a nearest neighboring site is smaller than the time required for spin fluctuations. Thus, the hole in its motion finds the spins nearly antiferromagnetically aligned which provide an almost linearly rising potential to which the hole is nearly bound.

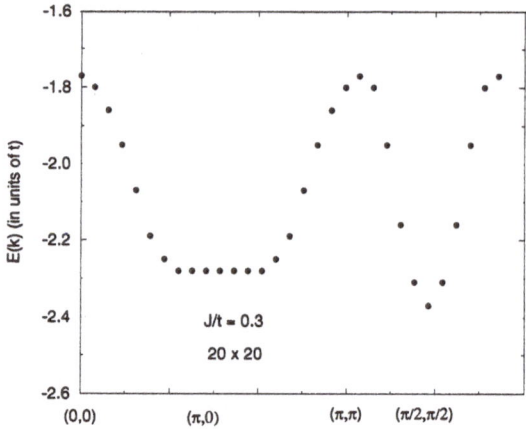

Fig. 3. The hole dispersion (Liu and Manousakis 1992).

The eigenstates of a linear potential correspond to the peaks of spectral function (Liu and Manousakis 1992). The actual problem of the hole in a fluctuating antiferromagnet has some subtle differences from this approximate picture. For example, spin fluctuations can repair the damage created by the hole motion and this gives rise to a hole bandwidth of the order of J.

The hole band has its minimum at $\mathbf{k} = (\pi/2, \pi/2)$ as can be seen from Fig. 3 and this is a general conclusion found by a number of other calculations. Our calculated dispersion has been compared to experimental results obtained by high resolution angle resolved photoemission on the insulating layered copper-oxide $Sr_2CuO_2Cl_2$ (see Wells et al., 1995). Reasonable agreement was found for the bandwidth (with no free parameter) and for the minimum of the band. There are important differences in the overall shape of the band along some crystallographic directions which may be removed if other small couplings are introduced such as a next nearest hole hopping term (Manousakis 1995).

In the GFMC calculation for a single hole we used as initial function the wave function (15) and the function

$$\Psi_G = exp\left(-\frac{1}{2}\sum_{i<j} u_{ij} s_i^z s_j^z\right), \qquad (18)$$

as guidance function.

In Fig. 4, BM show results of a transient estimation, for $t = 5J$ on a 4×4 lattice. The dashed line indicates the exact ground state energy (Barnes et al. 1992). To demonstrate the importance of the initial state, we compare the results obtained using the following two different trial states, with the same walker-population size: a) the state Ψ_T^I (results are shown by open circles in Fig. 4)

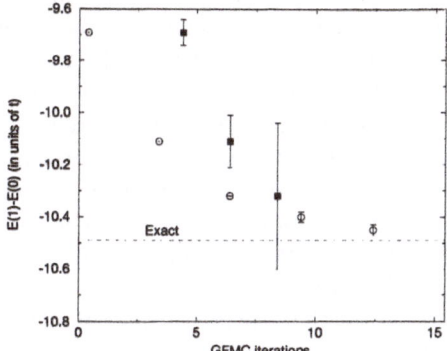

Fig. 4. Comparison of transient estimation for the energy of a single hole in a Heisenberg antiferromagnet with two different trial states (Boninsegni and Manousakis 1992b).

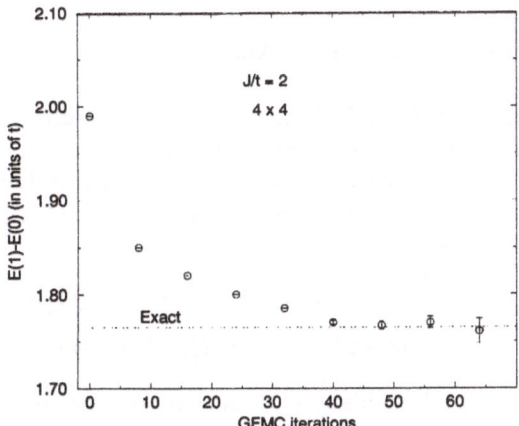

Fig. 5. Typical transient estimation for the energy of a single hole in a Heisenberg antiferromagnet (Boninsegni and Manousakis 1992b).

given by Eq. (15) with the optimal variational parameters and b) the state Ψ_T^{II} (results are shown by solid squares in Fig. 4), also given by Eq. (15) but with $f_a(\mathbf{k})$, $f_{aa'}(\mathbf{k}) = 0$, which corresponds to a state with no spin-hole correlations. The GFMC propagation of the trial state Ψ_T^I after n powers of the projection operator yields a much better ground state energy estimate; in addition, the size of the error bars is much smaller when the state Ψ_T^I is used, using the same number of walkers. As one can easily observe the error bars with the state Ψ_T^{II}

are large and they do not allow extrapolation to infinite power of the projection operator. Of course convergence can be achieved by increasing the number of walkers to reduce the error bars. However using Ψ_T^{II}, in order to obtain the same level of convergence to that obtained with the wave function Ψ_T^I, the number of walkers needs to be increased by a factor of order of at least 100 and the computer time by the same amount, while implementation of the more complex wave function Ψ_T^I increases the computer time by a factor of six only. From this demonstration it is clear that one needs both accurate trial and guiding states in order to control the size of the error bars. In addition, one needs clever extrapolation schemes to infinite power.

In Fig. 5 BM demonstrate that their GFMC calculation reproduces the results of exact diagonalization for a single hole in a 4×4 lattice. The GFMC calculation of BM, however, has been done on significantly larger size lattices than are possible by exact diagonalization. For details of this calculation for the single hole, the reader is referred to the work of BM (Boninsegni and Manousakis 1992a,Boninsegni and Manousakis 1992b). Generally speaking the calculation confirms the results obtained by the loop expansion calculation of Liu and Manousakis (Liu and Manousakis 1992).

6 Two holes in an infinite lattice

The GFMC calculation of BM (Boninsegni and Manousakis 1993) for two holes is based upon an initial state for two holes which is a generalization of the string-based variational state used in the single-hole calculation and on the same guidance function used in the calculation for a single hole. The correlated two-hole state is defined as follows:

$$|\Psi_T(\mathbf{Q})\rangle = \sum_{\mathbf{R},\mathbf{r},c}(-1)^{L(c)}e^{-i\mathbf{Q}\cdot(\mathbf{R}+\mathbf{r}/2)}\ \hat{F}_{\mathbf{Q}}(\mathbf{r})$$
$$\times exp(-\tfrac{1}{2}\sum_{i<j}u_{ij}\hat{s}_i^z\hat{s}_j^z)\ g(\mathbf{r})c_{\mathbf{R}\uparrow}c_{\mathbf{R}+\mathbf{r}\downarrow}|c\rangle, \tag{19}$$

where the sum runs over all lattice sites $\mathbf{R}$ and over all lattice spin configurations. Here $\mathbf{Q}$ is the total momentum of the state and $\mathbf{r}$ is the relative distance of the two holes. The "string" correlation operator is given by:

$$\hat{F}_{\mathbf{Q}}(\mathbf{r}) = 1 + \sum_a f_a(\mathbf{Q},\mathbf{r})\hat{\mathcal{P}}_a + \sum_{aa'} f_{aa'}(\mathbf{Q},\mathbf{r})\hat{\mathcal{P}}_{a'}\hat{\mathcal{P}}_a\ , \tag{20}$$

where $f_a(\mathbf{Q},\mathbf{r})$, $f_{aa'}(\mathbf{Q},\mathbf{r})$ are variational parameters and the operators $\hat{\mathcal{P}}_a$ were introduced in Eq. (16). The energy expectation value has been minimized by taking $g(\mathbf{r})$ to be non-zero for only n.n. and with $d_{x^2-y^2}$-wave spatial symmetry, i.e. $g(\pm\hat{x}) = -g(\pm\hat{y})$, which corresponds to a singlet state of the two holes. The guidance function Ψ_G used is given by Eq. (18).

In Fig. 6 the energy of two holes relative to the energy of the no-hole state as calculated by the GFMC calculation is compared to the exact (Barnes et al. 1992) on a 4×4 cluster to demonstrate the method. In Fig. 7 the two hole

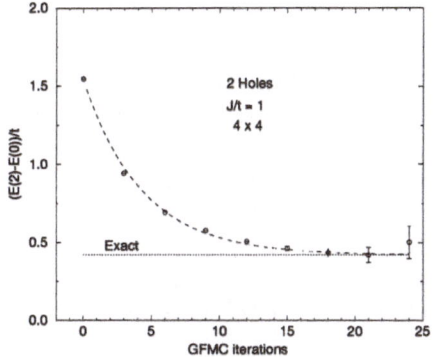

Fig. 6. Typical transient estimation for the energy of two holes in a Heisenberg anti-ferromagnet (Boninsegni and Manousakis 1993).

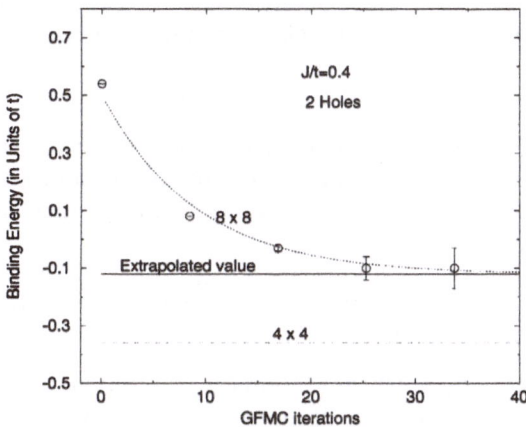

Fig. 7. Two-hole binding energy on an 8 × 8 lattice (Boninsegni and Manousakis 1993).

binding energy obtained on an 8 × 8 is compared to the corresponding binding on a 4 × 4 lattice for $J/t = 0.4$. This demonstrates the large overestimation of the binding in the small size systems which are accessible to exact diagonalization.

The reader is referred to the GFMC calculation of BM (Boninsegni and Manousakis 1993) for important details and conclusions from the study of two holes. The general conclusions are: a) There are significant finite-size effects in the calculation of small size systems obtained by straightforward diagonalization. b) There is a critical value of J/t of approximately 0.27 below which binding of two holes in a $d_{x^2-y^2}$ state is no longer possible.

7 Few electrons and finite density of holes/electrons

More recently, Hellberg and Manousakis (HM) have carried out GFMC calculations for the t-J model for a few electrons (Hellberg and Manousakis 1995, Hellberg and Manousakis 1996) and at finite densities of electrons (Hellberg and
Manousakis 1996, Hellberg and Manousakis 1997) for all interaction strengths.
A very general form for the trial state is written down which is restricted to be
a spin singlet with zero total momentum. The family of states which were used
can be summarized in the form:

$$|\Psi_T\rangle = \prod_{i<j,\sigma,\sigma'} f(\mathbf{r}_{i\sigma} - \mathbf{r}_{j\sigma'})P_N \prod_{\mathbf{k}}(u_{\mathbf{k}} + v_{\mathbf{k}}c^\dagger_{\mathbf{k}\uparrow}c^\dagger_{-\mathbf{k}\downarrow})|0\rangle, \qquad (21)$$

where P_N projects the state onto the subspace with fixed number of particles.
The Jastrow factor $f(\mathbf{r})$ correlates all pairs of particles, yielding a total spin
singlet correlated state. The hard-core constraint of the t-J model is satisfied
through $f(\mathbf{r} = \mathbf{0}) = 0$. It is clear that the projection of this state to a fixed number of particles, includes as a particular choice the correlated Slater determinant.
This corresponds to a specific choice of the pairing factors $u_{\mathbf{k}}$ and $v_{\mathbf{k}}$.

The GFMC requires a positive definite guiding function, which is taken to
be

$$\Psi_G = \max\{|\Psi_T|, c\Psi_B\}, \qquad (22)$$

where Ψ_B is a positive function such as the solution of the same Hamiltonian
for a two component boson system (say, up bosons and down bosons interacting
with the $t - J$ model Hamiltonian). An analytic approximation for the function
Ψ_B is the following:

$$\Psi_B = \prod_{i<j,\sigma,\sigma'} f_B(\mathbf{r}_{i\sigma} - \mathbf{r}_{j\sigma'}) \prod_{i,j} s_B(\mathbf{r}_{i\uparrow} - \mathbf{r}_{j\downarrow}). \qquad (23)$$

This function mimics as much as possible the fermion state without having
fermionic statistics. Since it is not important to guide with a spin singlet function, the additional spin-dependent Jastrow factor $s_B(\mathbf{r})$ is used in the guiding
function. The coefficient c is chosen to keep the same normalization for both
functions. The absolute magnitude of the trial state, is a good guiding function
except when the node falls on the lattice sites.

Careful choices of trial and guiding states can drastically reduce the statistical errors in the GFMC. Since the electrons on a given size lattice only occupy
a finite number of locations $\mathbf{r}$, the factors in (21) and (23) are varied independently at each distance or wave vector not related by symmetry (Hellberg and
Manousakis 1996). Thus, very general physical states afforded by the forms of
the trial and guiding functions were allowed. For example, on a 20×20 lattice,
there are 800 variational parameters for the Jastrow factors and the pairing functions. By taking into account all rotation and mirror symmetries for the Jastrow
factors and only the mirror symmetries about the axes and parity for the pairing
functions the number of free parameters are reduced to 172 for the trial state

and to 128 parameters for the guiding function. To optimize the parameters, the variance of the local energy was minimized. For sufficiently large J/t, the spin independent Jastrow factors $f(\mathbf{r})$ in (21) and (23) bind the electrons in the optimized state, while the pairing factors u_k and v_k and the spin-dependent Jastrow factor $s_G(\mathbf{r})$ provide the internal correlation of the bound system.

The statistical errors increase exponentially with increasing power n, so the GFMC output of all powers is used to extrapolate to infinite power. This is achieved by considering the calculated powers of $\hat{H}$ within the trial state $|\Psi_T\rangle$ as moments of the spectral function of a Green's function with respect to the trial state (Hellberg and Manousakis 1997). The poles of such a Green's function are at the exact eigenvalues of the eigenstates of $\hat{H}$ having non-zero overlap with $|\Psi_T\rangle$.

First, the boundary of phase separation (Ioffe and Larkin 1988, Emery et al., 1990) is determined. It is well known that, at large values of J/t and at any electron density $n_e \leq 1$, the system separates into a region with density $n_e = 1$ and an empty region. This is due to the following argument. An electron in a 2D Heisenberg antiferromagnet (as defined by the J term of Eq. (2) with the $1/4 n_i n_j$ term) has an average energy of $E_H = -1.1692J$ (Carlson 1989, Trivedi and Ceperley 1989, Runge 1992). A free electron has an energy of $-4t$ so when $E_H > -4t$, i.e $J/t < 3.42$, free electrons can evaporate, otherwise they stay bound to take advantage of the Heisenberg attraction. However, as soon as two electrons evaporate, they bind themselves in pairs which lowers the evaporation energy, thus the critical J for evaporation of pairs is $J_c = 3.4367t$ as found by HM (Hellberg and Manousakis 1995, Hellberg and Manousakis 1996) exactly. The small difference in these two values of J is due to the binding energy δ of an electron pair in an extended s-wave state (for $J > 2t$) and δ happens to be very small for this value of $J = J_c$. Larger clusters of electrons, such as quartets, never evaporate, because they become stable compared to pairs at $J > 5.21t$ and they are never stable compared to a Heisenberg infinite cluster (Hellberg and Manousakis 1995). These statements are demonstrated in Fig. 8.

At low density and for $J/t < 2$ the extended s-wave bound state is no longer stable. One might think that this corresponds to a paramagnetic state of free electrons. However, it can be shown (Chubukov and Kagan 1992, Chubukov 1993, Kagan and Rice 1994), that this state at a small finite density of electrons will always be unstable to p-wave pairing due to the Kohn-Luttinger effect. Thus, using low density expansion (T-matrix calculations) one can calculate the phase boundary of these phases which are expected to be accurate at low density. In Fig. 9 the phase diagram of the $t - J$ model at low electron density is shown as calculated by HM (Hellberg and Manousakis 1995) which we believe to be very accurate.

At higher densities HM have examined the question of phase separation using the GFMC method. They found that the phase separation line, at the critical electron sensity $n_c(J/t)$, extends for all interactions strengths. The calculated phase separation line $n_c(J/t)$ is shown in Fig. 10. In contrast to the earlier studies in which a vanishing inverse compressibility was used as a criterion for the onset

142

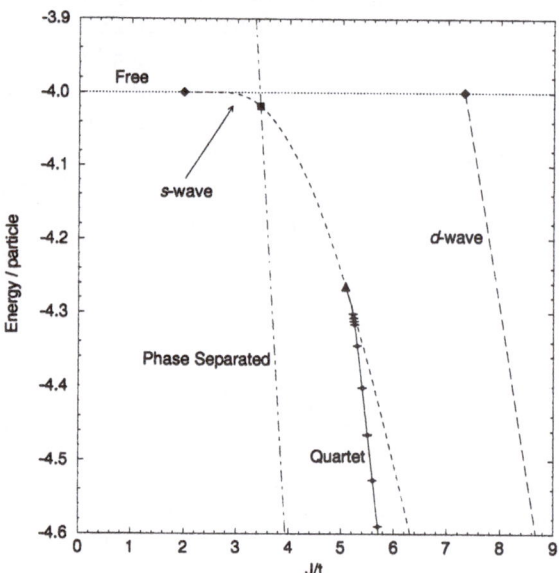

Fig. 8. The horizontal line is the energy of free electrons, while the dashed line labeled
s-wave is the energy per electron of two electrons paired in an extended s-wave state.
The solid line labeled quartet is the energy of four bound electrons. The dashed line
labeled d-wave is the energy of two-electrons bound in a d-wave state. The dashed
dotted line is the energy per electron of the Heisenberg antiferromagnet on a square
lattice. This figure is taken from the work of HM (Hellberg and Manousakis 1995).

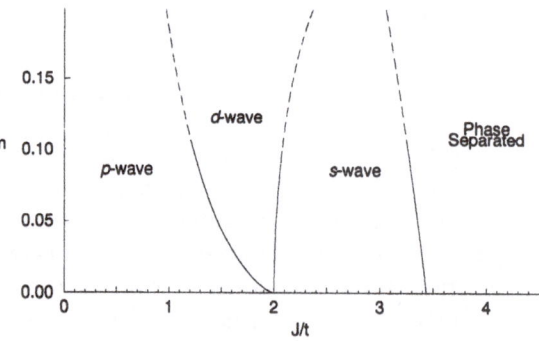

Fig. 9. The phase diagram of the $t - J$ model at low electron density. The phase
boundaries are accurate at low density, but they become increasingly inaccurate as the
density increases. This figure is taken from the work of HM (Hellberg and Manousakis
1995).

of phase separation, the phase separation line in the work of HM was determined using the Maxwell construction.

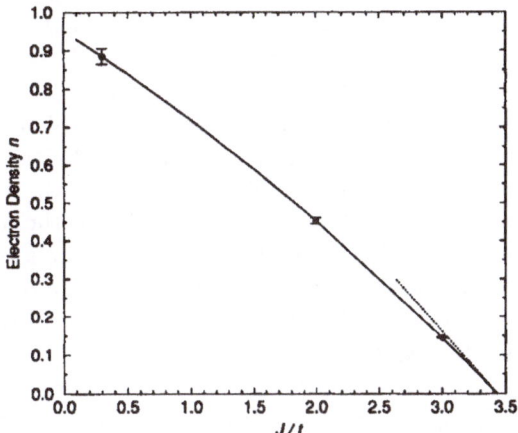

Fig. 10. The phase separation of the $t-J$ model as a function of electron density. In the limit of low hole density the model phase separates for all interaction strengths. This figure is taken from the work of Hellberg and Manousakis (Hellberg and Manousakis 1996, Hellberg and Manousakis 1997).

The phase diagram as calculated by HM differs substantially from the results of most recent studies (Luchini et al. 1991, Putikka et al. 1992a, Putikka et al. 1992b, Poilblanc 1995). In these studies the inverse compressibility was used to determine the phase separation line. However, the inverse compressibility, in a first order phase transition, vanishes discontinuously and in fact it is always finite in the uniform phase for a 2D system. In the region where one has phase separation, in an infinite system the inverse compressibility vanishes. However, the energy of a finite system in the phase separated region suffers strong finite size effects. The reason is that the coexistence of the two phases requires a surface and in a small size system the surface to bulk ratio is not small. This brings in important finite-size effects. As a result below a value of $J/t \sim 1$, the compressibility of small size systems available to exact diagonalization never diverges. This was interpreted that the phase separation line dies off below some value of J/t. Thus, the conclusion of these studies was that the physical region of the model is outside the phase separated region. This was the conclusion of small size diagonalization (Poilblanc 1995) and of high temperature series expansion studies (Luchini et al. 1991, Putikka et al. 1992a, Putikka et al. 1992b). The latter studies also suffer from similar problems, because from a high temperature series expansion one necessarily needs to extrapolate from the high temperature phase.

At small electron density the phase separation line is given by $J_c(n_e \to 0)/t = J_c(0)/t - \alpha n_e$ where $J_c(0)/t \simeq 3.4367$ and $\alpha \simeq \pi/E_H = 2.887$ where E_H is the energy of the J term of the $t - J$ Hamiltonian in the limit of $n_e = 1$. This line is obtained by comparing the kinetic energy of small density of free electrons which are paired in singlet states having binding energy with approximately negligible density dependence. The binding energy per electron at $J \sim J_c(0)$ is of order of $0.1t$ while the kinetic energy per electron is $-4t + \pi n_e t$. The line $J_c(n_e \to 0)$ is shown in Fig. 10 as a dotted line.

Phase separation all the way in the physical region of the model was in the past proposed by Emery et al. (Emery et al., 1990) and by Marder et al. (Marder et al. 1990). The bulk of the later studies (Luchini et al. 1991, Putikka et al. 1992a, Putikka et al. 1992b, Poilblanc 1995) questioned the foundation of these works; thus, such later studies created the belief that the $t - J$ does not phase separate in the physical region. As was shown by the more reliable calculation of HM, however, for any value of J/t there is a critical value of the electron density $n_c(J/t)$ such that for $n > n_c$ there is phase separation.

Finally, HM have used their knowledge about the various regions of the phase diagram to derive a proposal for the full phase diagram of the model. First of all one knows with good degree of confidence the phase separation line obtained from the work of HM. It is shown by the solid line in Fig. 11. One also knows what happens at low electron density. The lines drawn as solid lines are phase boundaries known with rather good degree of confidence. The boundaries of these instabilities at higher density are only drawn with dashed lines indicating that they are more speculative. These lines are calculated by pursuing the calculation of Kagan and Rice (Kagan and Rice 1994) which is valid at low density. These phases may continue at higher densities but one cannot trust the precise location of the lines as obtained by low density expansion. The phase boundary between d- and p-wave is expected to meet the phase separation boundary at $J^* \simeq 0.27t$. This value of J was found to be the minimum value of J for which a two-hole d-wave bound state was stabilized (Boninsegni and Manousakis 1993). Now, having the evidence of phase separation, one interprets the two hole binding as evidence of phase separation in a state of d-wave symmetry. The calculated phase boundary using the expressions from the low density expansion gives $J/t \sim 0.5$ as $n_e \to 1$; however, this line has been modified so that to end somewhere close to J^*. For $J < 0.1t$ there is a ferromagnetic instability at small hole dopings (Ioffe and Larkin 1988, Emery et al., 1990, Marder et al. 1990, Putikka et al. 1992b) and this has been also sketched in the phase diagram. In the unpolarized region a phase with p-wave pairing, a continuation of the low density phase, is compatible with ferromagnetic correlations near the ferromagnetic instability.

The calculation of HM cannot exclude a reentrance of a homogeneous phase at very low hole density separated from the low electron density uniform phase by a phase separation. This phase diagram may be sensitive to other terms that one may add to the $t - J$ Hamiltonian. An example is the long-range Coulomb repulsion, namely the fact that the two phases when separated have different macroscopic charges which will prevent the phase separation. The local

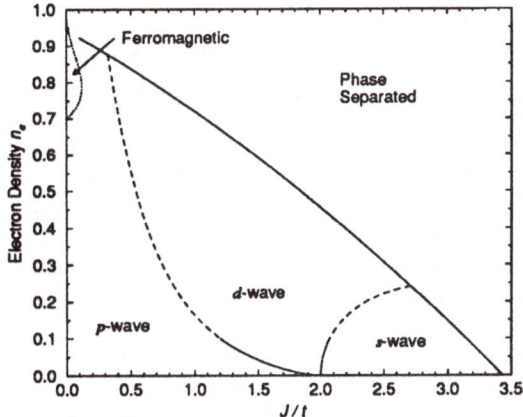

Fig. 11. The phase diagram of the $t - J$ model as a function of electron density. The solid lines correspond to reliable calculations of the phase boundaries by HM (Hellberg and Manousakis 1996, Hellberg and Manousakis 1997). The dashed lines are estimates and correspond to the simplest extension of our knowledge using the requirement of connectivity of the phase diagram.

tendency for phase separation however could have consequences for the dynamics of the more complete Hamiltonian. It is interesting to find out what are the consequences of this tendency in its competition with the Coulomb forces. It might lead to creation of stable clusters of definite sizes or might give rise to charged stripes as those found by Tranquada et al. (Tranquada 1995,Tranquada 1996) or other scenarios such as those considered by Emery et al. (Emery et al. 1993-96).

8 Acknowledgements

This work was supported by the Office of Naval Research under Grant No. N00014-93-1-0189.

References

Barnes, T., Jacobs, A. E., Kovarik, M. D., Macready, W. G., 1992 Phys. Rev. **B 45**, 256.

Boninsegni, M., and Manousakis, E., 1992a, Phys. Rev. B **45**, 4877.

Boninsegni, M., and Manousakis, E., 1992b, Phys. Rev. B **46**, 560.

Boninsegni, M., and Manousakis, E., 1993, Phys. Rev. B **47**, 11897.

Carlson, J., 1989, Phys. Rev. **B 40**, 846.

Chubukov, A. V., Kagan, M. Y., 1992, in *Physical Phenomena at High Magnetic Fields*, eds E. Manousakis et al. pg. 239, Addison-Wesley.

146

Chubukov, A. V., 1993, Phys. Rev. B **48**, 1097.

Dagotto, E., 1994, Rev. Mod. Phys. **66**, 763.

Emery, V. J., Kivelson, S. A., and Lin, H. Q., 1990, Phys. Rev. Lett. **64**, 475.

Emery V. J., and Kivelson, S. A., 1993, Physica C **209**, 597. Emery V. J. and Kivelson, S. A., 1994, Physica C **235-40**, 189. Emery V. J. and Kivelson, S. A., 1995, Phys. Rev. Lett. **74**, 3253. Salkola, M.I., Emery V. J., and Kivelson, S. A., 1996, **77**, 155.

Hellberg, C. S., and Mele, E. J., 1993, Phys. Rev. B **48**, 646.

Hellberg C. S., and Manousakis, E., 1995, Phys. Rev. B **52**, 4639.

Hellberg, C. S., and Manousakis, E., 1996, in *Physical Phenomena at High Magnetic Fields - II*, edited by Z. Fisk et al. (World Scientific, Singapore).

Hellberg, C. S., and Manousakis, E., 1996, preprint cond-mat 9611195, submitted to Phys. Rev. Lett.

Hellberg, C. S., and Manousakis, E. 1997, to be submitted in Phys. Rev. B.

Ioffe L. B., and Larkin, A. I., 1988, Phys. Rev. B **37**, 5730.

Kagan, M. Y., and Rice, T. M., 1994, J. Phys. Cond. Mat. **6**, 3771.

Kalos, M. H., 1962, Phys. Rev. **128**, 1791.

Kalos, M. H., 1966, J. Comput. Phys. **1**, 257.

Kalos, M. H., 1981, Lee, M.A., Whitlock P. A., and Chester G. V., Phys. Rev. B **24**, 115.

Liu, Z., and Manousakis, E., 1992, Phys. Rev. B **45**, 2425.

Luchini, M. U., Ogata, M., Putikka, W. O., and Rice, T. M., 1991, Physica C **185-189**, 141.

Marder, M., Papanicolaou, N., and Psaltakis, G. C., 1990, Phys. Rev. B **41**, 6920.

Manousakis, E., 1991, Rev. Mod. Phys. **63**, 1.

Manousakis, E., 1995, in *Electronic, Optoelectronic and Magnetic Thin Films*, eds J. M. Marschall, N. Kirov and A. Varvek, Research Studies Press LTD.

Negele, J. W. and Orland, H. , 1988, *Quantum Many-Particle Systems*, Frontiers in Physics, Addison-Wesley.

Poilblanc, D., 1995, Phys. Rev. B **52**, 9201, and references there in.

Putikka, W. O., Luchini, M. U., and Rice, T. M., 1992, Phys. Rev. Lett. **68**, 538.

Putikka, W. O., Luchini, M. U., and Ogata, M., 1992, Phys. Rev. Lett. **69**, 2288.

Runge K. J., 1992, Phys. Rev. B **45**, 12292.

Tranquada, J. M., *et al.*, 1995, Nature, **375**, 561.

Tranquada, J. M., Axe, J. D., Ichikawa, N., Moodenbaugh, A. R., Nakamura, Y. and Uchida, S. , 1996, preprint, cond-mat 9608048.

Trivedi, N. and Ceperley D. M., 1989, Phys. Rev. B **40**, 2747.

Wells, B. O., Shen, Z.-X., Matsuura, A., King, D. M., Kastner, M. A., Greven, M. and Birgeneau, R. J., 1995, Phys. Rev. Lett. **74**, 964.

Yokoyama H. and Ogata, M., 1996, to appear in J. Phys. Soc. Japan.

Dynamics of Non-Planar Vortices in the Classical 2D Anisotropic Heisenberg Model at Finite Temperatures

T. Kamppeter[1], F. G. Mertens[1], Angel Sánchez[2], N. Gronbech-Jensen[3], A. R. Bishop[3] and F. Domínguez-Adame[4]

[1] Physikalisches Institut, Universität Bayreuth, D-95440 Bayreuth, Germany
[2] Departamento de Matemáticas and Grupo Interdisciplinar de Sistemas Complicados, Universidad Carlos III de Madrid, c/ Butarque 15, E-28911 Leganés, Spain
[3] Theoretical Division, Los Alamos National Laboratory, MS B262, Los Alamos, New Mexico 87545, USA
[4] Departamento de Física de Materiales and Grupo Interdisciplinar de Sistemas Complicados, Facultad de Físicas, Universidad Complutense, E-28040 Madrid, Spain

Abstract. The 2-dimensional anisotropic Heisenberg model with XY- or easy-plane symmetry bears non-planar vortices which exhibit a localized structure of the z-components of the spins around the vortex center. In order to study the dynamics of these vortices under thermal fluctuations we use the Landau-Lifshitz equation and add white noise and Gilbert damping. Using a collective variable theory we derive an equation of motion with stochastic forces which are shown to represent white noise with an effective diffusion constant. We compare the results with Langevin dynamics simulations for the Landau-Lifshitz equation and find three temperature regimes: For low temperatures the dynamics is described by a 3rd-order equation of motion, for intermediate temperatures by a 1st-order equation. For higher temperatures, but still below the Kosterlitz-Thouless transition temperature, the spontaneous appearance of vortex-antivortex pairs does not allow a single-particle description.

1 Introduction

Very recently Mertens et al. [1] developed a collective variable theory for non-linear coherent excitations in classical systems with arbitrary Hamiltonians. In this theory the dynamics of a single excitation is governed by a hierarchy of equations of motion for the excitation center $\mathbf{X}(t)$. The type of the excitation determines on which levels the hierarchy can be truncated consistently: So-called gyrotropic excitations are governed by odd-order equations and thus do not have Newtonian dynamics, e. g. Galileo's law is not valid. Non-gyrotropic excitations are described by even-order equations, i. e. by Newton's equation in the first approximation. Examples of the latter case are kinks in 1-dimensional models and planar vortices in the 2-dimensional anisotropic Heisenberg model with XY- or easy-plane symmetry. The non-planar vortices of this model represent the simplest gyrotropic example.

The word "gyrotropic" was coined because these vortices are subject to a gyrocoupling force $\dot{\mathbf{X}} \times \mathbf{G}$ [2], [3], which is formally equivalent to the Lorentz force. However, the gyrocoupling vector [4]

$$\mathbf{G} = 2\pi q p \mathbf{e}_z \qquad (1)$$

does not represent an external field but it is an intrinsic quantity which is produced by a well-localized structure of the spin components S_z around the vortex center ($\mathbf{e}_z$ is the unit vector perpendicular to the xy-plane of the magnetic ions which carry the spin $\mathbf{S}$); q is the vorticity and p is a second topological charge, defined as the value of S_z at the vortex center. The word "non-planar vortex" was chosen because of this S_z-structure (Planar vortices have $p = 0$ and are therefore not subject to a gyrocoupling force).

In this paper we study how thermal fluctuations influence the dynamics of nonlinear collective excitations in classical spin systems, specifically focusing on non-planar vortices. We begin with the Landau-Lifshitz equation, which describes the microscopic spin dynamics, and add noise $\eta^m(t)$ and damping (required to fulfill the dissipation-fluctuation theorem):

$$\frac{d\mathbf{S}^m}{dt} = -\mathbf{S}^m \times \frac{\partial H}{\partial \mathbf{S}^m} - \epsilon \mathbf{S}^m \times \frac{d\mathbf{S}^m}{dt} + \eta^m \quad . \qquad (2)$$

Here $\mathbf{S}^m$ is the spin vector at lattice site m, H is the Hamiltonian, and ϵ is the damping parameter. Following Refs. [2], [3] and [4] we have chosen Gilbert damping[1], chiefly because it is isotropic, in contrast to the Landau-Lifshitz damping [5]. For simplicity, we have added in Eq. (2) three noise components to the deterministic equations, although these do not have the form of Langevin equations (all components of $d\mathbf{S}/dt$ appear in each equation). In Sec. 3 we will introduce noise in the proper way, which will eventually lead to an equation similar to (2).

The Hamiltonian of our model is

$$H = -J \sum_{<m,n>} \left[S_x^m S_x^n + S_y^m S_y^n + (1 - \delta) S_z^m S_z^n \right] \qquad (3)$$

with $0 < \delta \leq 1$, and $< m, n >$ labels nearest neighbors of a square lattice.

In this work, we want to extend the results in [1] to finite temperatures. Our approach to this problem is both analytical and numerical: We first derive equations of motion for the vortex center $\mathbf{X}(t)$ which contain damping as well as stochastic forces. Afterwards we compare with Langevin dynamics simulations for our model, i. e., with results from numerical integration of (2).

We proceed in four steps: (a) We take (2) without damping and noise and reproduce the equation of motion of Ref. [1], which started from the Hamilton equations, however. (b) We include the damping term and find out how it affects the trajectories. (c) We further include the noise and calculate the mean value and variance of the resulting stochastic forces exerted on the vortex. Our main result in this regard is the fact that these forces are given by white noise with

---

[1] The sign of our damping term differs from [2]–[5] because we work with spins while these authors deal with magnetisations.

an effective diffusion constant. (d) We compare with the simulations and find several temperature regimes in which the dynamics differs qualitatively. We will see that for two regimes the dynamics can be described by equations of motion from two different levels of the above mentioned hierarchy, namely by the 3rd-order equation in the low-temperature region and the 1st-order equation in the intermediate regime. For higher temperatures the spontaneous appearence of many vortex-antivortex pairs does not permit a single-particle description.

2 Equations of Motion with and without Damping

We work in the continuum limit and replace $\mathbf{S}^m(t)$ by the spin field $\mathbf{S}(\mathbf{r}, t)$. For steady state motion, when the shape of the excitation is rigid, Thiele [2], [3] used the travelling wave ansatz $\mathbf{S}(\mathbf{r}, t) = \mathbf{S}(\mathbf{r} - \mathbf{X}(t))$ with constant velocity $\dot{\mathbf{X}}$ and derived the 1st-order equation of motion $\mathbf{G} \times \dot{\mathbf{X}} + \mathbf{F} = 0$, where $\mathbf{F}$ is a static force, due to either an external field or the interactions with other excitations.

However, on an arbitrary trajectory the shape of a collective excitation naturally depends on the velocity $\dot{\mathbf{X}}$ and, as shown in [1], in general also on higher order derivatives of $\mathbf{X}(t)$. The corresponding generalized travelling wave ansatz is

$$\mathbf{S}(\mathbf{r}, t) = \mathbf{S}(\mathbf{r} - \mathbf{X}, \dot{\mathbf{X}}, \ddot{\mathbf{X}}, \dots, \mathbf{X}^{(n)}) \ , \tag{4}$$

which yields an $(n + 1)^{\text{th}}$-order differential equation for $\mathbf{X}(t)$. For gyrotropic excitations only odd-order equations are relevant. In the case of the non-planar vortices it turned out that the 3rd-order equation is completely sufficient to describe accurately all simulations[2] which were made in [1]. Therefore in this paper we use the ansatz (4) with $n = 2$.

We want to get an equation for force densities which will lead to forces after an integration over the system. Therefore we perform the following operations with (2), leaving out damping and noise for the moment

$$\mathbf{S}\left(\frac{\partial \mathbf{S}}{\partial X_i} \times \frac{d\mathbf{S}}{dt}\right) = -\mathbf{S}\left(\frac{\partial \mathbf{S}}{\partial X_i} \times \left[\mathbf{S} \times \frac{\delta H}{\delta \mathbf{S}}\right]\right) = -S^2 \frac{\delta H}{\delta \mathbf{S}} \frac{\partial \mathbf{S}}{\partial X_i} = -S^2 \frac{\partial \mathcal{H}}{\partial X_i} \tag{5}$$

with $i = 1, 2$ in the case of our 2D system. $\mathcal{H}$ is the Hamiltonian density. According to our ansatz we insert on the l.h.s.

$$\frac{d\mathbf{S}}{dt} = \frac{\partial \mathbf{S}}{\partial X_j} \dot{X}_j + \frac{\partial \mathbf{S}}{\partial \dot{X}_j} \ddot{X}_j + \frac{\partial \mathbf{S}}{\partial \ddot{X}_j} \dddot{X}_j \ , \tag{6}$$

integrate over $\mathbf{r}$ and divide by S^2. In this way we get the same 3rd-order equation as that obtained in Ref. [1] which started from the Hamilton equations:

$$\mathbf{A}\dddot{\mathbf{X}} + \mathbf{M}\ddot{\mathbf{X}} + \mathbf{G}\dot{\mathbf{X}} = \mathbf{F} \tag{7}$$

---

[2] The only exception was a certain combination of initial and boundary conditions which was specially designed in order to observe the effects predicted by the 5th-order equation [1].

with the force

$$F_i = -\int d^2r \frac{\partial \mathcal{H}}{\partial X_i} \quad ,$$

(8)

the gyrotensor

$$G_{ij} = S^{-2}\int d^2r\, \mathbf{S}\frac{\partial \mathbf{S}}{\partial X_i}\times\frac{\partial \mathbf{S}}{\partial X_j} = \int d^2r\left\{\frac{\partial\phi}{\partial X_i}\frac{\partial\psi}{\partial X_j} - \frac{\partial\phi}{\partial X_j}\frac{\partial\psi}{\partial X_i}\right\} \quad ,$$

(9)

the mass tensor

$$M_{ij} = S^{-2}\int d^2r\, \mathbf{S}\frac{\partial \mathbf{S}}{\partial X_i}\times\frac{\partial \mathbf{S}}{\partial \dot{X}_j} = \int d^2r\left\{\frac{\partial\phi}{\partial X_i}\frac{\partial\psi}{\partial \dot{X}_j} - \frac{\partial\phi}{\partial \dot{X}_j}\frac{\partial\psi}{\partial X_i}\right\} \quad ,$$

(10)

and the 3rd-order gyrotensor

$$A_{ij} = S^{-2}\int d^2r\, \mathbf{S}\frac{\partial \mathbf{S}}{\partial X_i}\times\frac{\partial \mathbf{S}}{\partial \ddot{X}_j} = \int d^2r\left\{\frac{\partial\phi}{\partial X_i}\frac{\partial\psi}{\partial \ddot{X}_j} - \frac{\partial\phi}{\partial \ddot{X}_j}\frac{\partial\psi}{\partial X_i}\right\} \quad .$$

(11)

The classical spin is constrained to have a fixed magnitude which we set to unity. So we get the expressions on the right using the canonical fields $\phi = \arctan(S_y/S_x)$ and $\psi = S_z$ for the spin field:

$$\mathbf{S} = \sqrt{1-\psi^2}\cos\phi\,\mathbf{e}_x + \sqrt{1-\psi^2}\sin\phi\,\mathbf{e}_y + \psi\,\mathbf{e}_z \quad .$$

(12)

Now we consider the Gilbert damping term in (2) and perform the same operations as above:

$$\epsilon\mathbf{S}\left[\frac{\partial\mathbf{S}}{\partial X_i}\times\left(\mathbf{S}\times\frac{d\mathbf{S}}{dt}\right)\right] = \epsilon S^2 \frac{\partial\mathbf{S}}{\partial X_i}\frac{d\mathbf{S}}{dt}$$

$$= \epsilon S^2\left[\frac{\partial\mathbf{S}}{\partial X_i}\frac{\partial\mathbf{S}}{\partial X_j}\dot{X}_j + \frac{\partial\mathbf{S}}{\partial X_i}\frac{\partial\mathbf{S}}{\partial \dot{X}_j}\ddot{X}_j + \frac{\partial\mathbf{S}}{\partial X_i}\frac{\partial\mathbf{S}}{\partial \ddot{X}_j}\dddot{X}_j\right] \quad .$$

(13)

An integration over $\mathbf{r}$ gives three terms which can be combined with the three terms on the l.h.s. of (7), i. e. the damping appears in every order

$$(\mathbf{A}+\mathbf{a})\dddot{\mathbf{X}} + (\mathbf{M}+\mathbf{m})\ddot{\mathbf{X}} + (\mathbf{G}+\mathbf{g})\dot{\mathbf{X}} = \hat{\mathbf{A}}\dddot{\mathbf{X}} + \hat{\mathbf{M}}\ddot{\mathbf{X}} + \hat{\mathbf{G}}\dot{\mathbf{X}} = \mathbf{F} \quad .$$

(14)

The damping tensors are

$$g_{ij} = \epsilon\int d^2r\,\frac{\partial\mathbf{S}}{\partial X_i}\frac{\partial\mathbf{S}}{\partial X_j} = \epsilon\int d^2r\left\{(1-\psi^2)\frac{\partial\phi}{\partial X_i}\frac{\partial\phi}{\partial X_j} + \frac{1}{1-\psi^2}\frac{\partial\psi}{\partial X_j}\frac{\partial\psi}{\partial X_i}\right\} \quad ,$$

(15)

$$m_{ij} = \epsilon\int d^2r\,\frac{\partial\mathbf{S}}{\partial X_i}\frac{\partial\mathbf{S}}{\partial \dot{X}_j} = \epsilon\int d^2r\left\{(1-\psi^2)\frac{\partial\phi}{\partial X_i}\frac{\partial\phi}{\partial \dot{X}_j} + \frac{1}{1-\psi^2}\frac{\partial\psi}{\partial \dot{X}_j}\frac{\partial\psi}{\partial X_i}\right\} \quad ,$$

(16)

$$a_{ij} = \epsilon\int d^2r\,\frac{\partial\mathbf{S}}{\partial X_i}\frac{\partial\mathbf{S}}{\partial \ddot{X}_j} = \epsilon\int d^2r\left\{(1-\psi^2)\frac{\partial\phi}{\partial X_i}\frac{\partial\phi}{\partial \ddot{X}_j} + \frac{1}{1-\psi^2}\frac{\partial\psi}{\partial \ddot{X}_j}\frac{\partial\psi}{\partial X_i}\right\} \quad .$$

(17)

The 1st-order part of (14) was already derived by Thiele [2].

An explicit calculation of all the tensor components is possible only if the *dynamic* structure of the collective excitation is known. The Hamiltonian density derived from (3) reads [8]

$$\mathcal{H} = \frac{JS^2}{2} \left\{ (1 - \psi^2)(\nabla\phi)^2 + \delta \left[4\psi^2 - (\nabla\psi)^2 \right] + \frac{1}{1 - \psi^2}(\nabla\psi)^2 \right\} . \quad (18)$$

In [1] the Hamilton equations were considered for a non-planar vortex in the center of a circular system with free boundary conditions. The vortex structure is complicated in an inner region $0 \leq r \leq a_c \approx 3r_v$, where

$$r_v = \frac{1}{2}\sqrt{\frac{1-\delta}{\delta}} \quad (19)$$

characterizes the vortex core [8]. δ is the anisotropy parameter in (3). Non-planar vortices are stable for $0 < \delta < 0.28$ for a square lattice [8]. We use $\delta = 0.1$ for our simulations. However, the inner region contributes very little to the integrals in (10), (11) and (15)–(17). Except for (9), the dominant contributions stem from the outer region $a_c \leq r \leq L$, where L is the system radius. Here the vortex has the following dynamic structure, which is also confirmed by simulations [1]:

$$\phi = \phi_0 + \phi_1 + \phi_2 , \quad \psi = \psi_0 + \psi_1 + \psi_2 \quad (20)$$

with

$$\phi_0 = q \tan^{-1}\frac{x_2}{x_1} , \qquad \psi_0 \sim p\sqrt{\frac{r_v}{r}} e^{-r/r_v} , \quad (21)$$

$$\phi_1 = p(x_1\dot{X}_1 + x_2\dot{X}_2) , \qquad \psi_1 = \frac{q}{4\delta r^2}(x_2\dot{X}_1 - x_1\dot{X}_2) , \quad (22)$$

$$\phi_2 = \frac{q}{8\delta} \ln r(x_2\ddot{X}_1 - x_1\ddot{X}_2) , \quad \psi_2 = \frac{p}{4\delta}(x_1\ddot{X}_1 + x_2\ddot{X}_2) . \quad (23)$$

Straightforward integrations then yield

$$G_{ij} = G\epsilon_{ij} , \quad G = 2\pi pq , \quad (24)$$

$$M_{ij} = M\delta_{ij} , \quad M = \frac{\pi q^2}{4\delta} \ln\frac{L}{a_c} + C_M , \quad (25)$$

$$A_{ij} = A\epsilon_{ij} , \quad A = \frac{G}{16\delta}\left(L^2 - a_c^2\right) + C_A , \quad (26)$$

$$g_{ij} = g\delta_{ij} , \quad g = \epsilon\pi q^2 \ln\frac{L}{a_c} + C_g , \quad (27)$$

$$m_{ij} = m\epsilon_{ij} , \quad m = \epsilon\frac{G}{4}\left(L^2 - a_c^2\right) + C_m , \quad (28)$$

$$a_{ij} = a\delta_{ij} , \quad a = \epsilon\frac{\pi q^2}{8\delta}\left[\frac{1}{2}\left(L^2 \ln L - a_c^2 \ln a_c\right) - \frac{1}{4}\left(L^2 - a_c^2\right)\right] + C_a , \quad (29)$$

where δ_{ij} is the 2D unit matrix, ϵ_{ij} is the antisymmetric tensor, and the different constants C are the contributions from the inner region of the vortex. We see that in every odd order of (14) a symmetric damping matrix is combined with an antisymmetric normal (non-damping) matrix, and vice versa for the even orders. Moreover the size dependence of the nth-order damping components is the same as that of the $(n+1)^{\mathrm{th}}$-order normal components. The 1st-order damping elements (27) were already evaluated in Refs. [3] and [7].

For the solution of the equation of motion (14) we proceed as in Ref. [1]: We consider small displacements $\mathbf{x}$ from a mean trajectory $\mathbf{X}^0$, on which the vortex is driven by $\mathbf{F}$

$$\mathbf{X}(t) = \mathbf{X}^0(t) + \mathbf{x}(t) . \tag{30}$$

In view of our simulations we consider a situation where the force is always pointing in the X_1-direction and expand to 1st order around $X_1(0) = R_0$ (This is justified because in our simulations F_0, and even more F_0', is very small):

$$F = F_0 + F_0' x_1 . \tag{31}$$

For $X_i^0(t)$ we get two coupled linear 3rd-order equations. For the initial conditions $X_1^0(0) = R_0$, $X_2^0(0) = 0$ the solutions are

$$X_1^0 = R_0 + \frac{F_0}{F_0'}(\mathrm{e}^{t/\tau} - 1) , \quad X_2^0 = \frac{G}{g}\frac{F_0}{F_0'}(\mathrm{e}^{t/\tau} - 1) , \tag{32}$$

where τ is determined by a cubic equation. The mean trajectory is a straight line $X_2^0 = G/g(X_1^0 - R_0)$, which slightly deviates from the X_2-axis. The angle g/G is small because $g \sim \epsilon$, where we choose small damping constants ϵ in the simulations. As τ is of the order of $G^2/(gF_0')$ it is very large, in fact much larger than our integration times. Therefore one can expand (32) and one get a constant velocity on the mean trajectory: $\dot{X}_1^0 = gF_0/G^2$, $\dot{X}_2^0 = F_0/G$.

The motion around the mean trajectory is obtained by solving two coupled linear 3rd-order equations for the displacements $\mathbf{x}(t)$ using the ansatz

$$x_i = x_i^0 \mathrm{e}^{-(\beta - i\omega)t} . \tag{33}$$

We get

$$\beta - i\omega = \frac{\pm iM + m \pm \sqrt{(\pm iM + m)^2 - 4(A \pm ia)(G \pm ig)}}{2(A \pm ia)} , \tag{34}$$

with amplitude ratios $\kappa = x_2^0/x_1^0 = \pm 1$ and phase differences $\pm\pi/2$, where we have set $F_0' = 0$ for simplicity. With $F_0' \neq 0$ Eq. (34) becomes even more complicated and $|\kappa| \neq 1$. The separation of real and imaginary parts leads to cumbersome formulas. Therefore we compute the frequencies $\omega_{1,2}$ and the relaxation constants $\beta_{1,2}$ as a function of the parameters ϵ and L; we choose $q = p = 1$ for the charges and $\delta = 0.1$ for the anisotropy. The a_c-dependent parts in (25)-(29)

can be combined with the constants C_M etc.; the combined constants can be neglected for large systems.

As $\omega_{1,2}$ turn out to be very close to each other we have plotted instead of $\omega_{1,2}$ the mean and the difference, respectively,

$$\omega_c = \sqrt{\omega_1 \omega_2} \ , \quad \Delta\omega = \omega_2 - \omega_1 \ , \tag{35}$$

as a function of ϵ for fixed L (Fig. 1) and vice versa (Fig. 2).

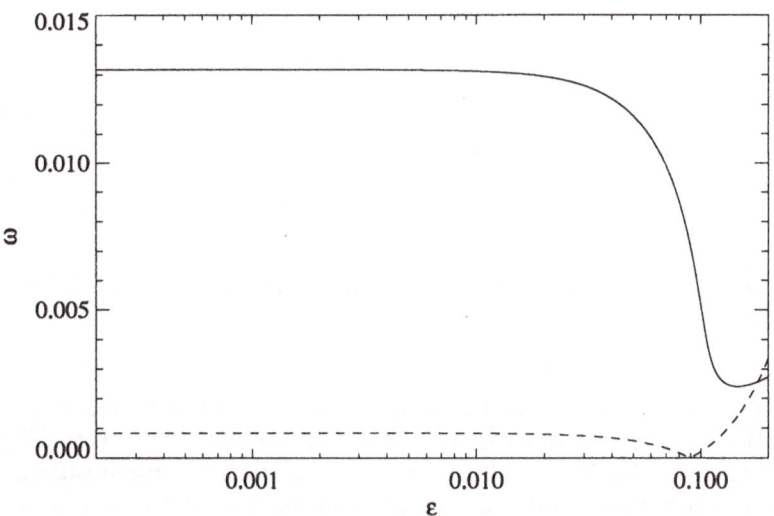

Fig. 1. Mean frequency ω_c and frequency difference $\Delta\omega$ vs. damping constant ϵ for $L = 24$, ——— : ω_c, − − − : $\Delta\omega$

In the simulations the purpose of the damping is to dissipate the energy which is supplied to the system by the noise. Therefore we must know the range of ϵ (for a given system size) in which the frequencies are not influenced by the damping. Figs. 1 and 2 reveal that this range is defined by the condition

$$\epsilon L \ll 5 \ . \tag{36}$$

The relaxation constants $\beta_{1,2}$ are nearly equal and the mean value is $\beta_c = \epsilon/8$.

For the above range ω_c and $\Delta\omega$ are related to the parameters G, M, and A in a very simple way [1]

$$\omega_c = \sqrt{\frac{G}{A}} \sim \frac{1}{L} \ , \quad \Delta\omega = \frac{M}{A} \sim \frac{\ln L}{L^2} \ . \tag{37}$$

The size dependences are demonstrated in Fig. 2.

154

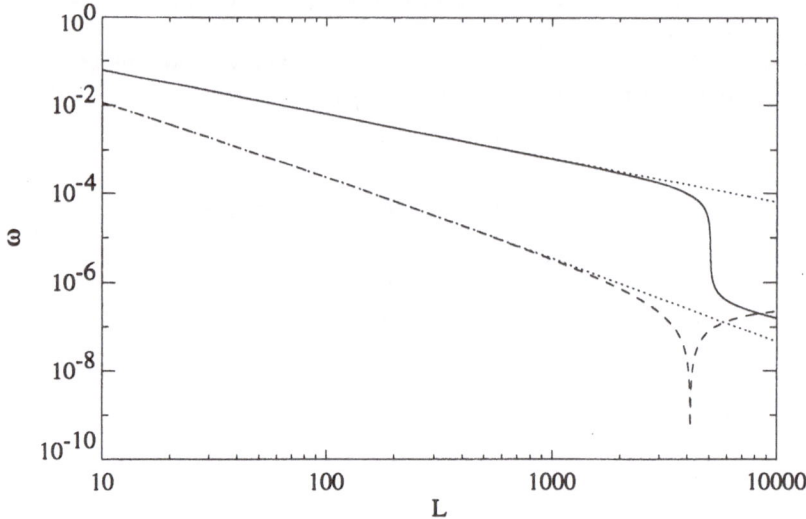

Fig. 2. Mean frequency ω_c and frequency difference $\Delta\omega$ vs. system size L for $\epsilon = 10^{-3}$,
————: ω_c,   – – –: $\Delta\omega$,   ······: $\epsilon = 0$

Finally we briefly discuss the shape of the trajectories. We first consider the motion in a frame which is moving along the mean trajectory $\mathbf{X}^0(t)$: The general solutions for the displacements $x_i(t)$ are linear superpositions of (33) with $\omega_{1,2}$. Both $x_i(t)$ exhibit a very pronounced beat because $\omega_{1,2}$ are nearly equal. The orbits $x_1(x_2)$ are Lissajous curves, which can look very intricate for certain parameter ranges. We go into the laboratory frame by adding $\mathbf{X}^0(t)$. Without the splitting of $\omega_{1,2}$ we would get a cycloid. Due to the splitting we finally get a superposition of two cycloids, which are slightly damped because of $\beta_{1,2}$.

3   Equations of Motion with Stochastic Forces

We want to introduce noise into the Landau-Lifshitz equation with damping

$$\frac{d\mathbf{S}}{dt} = -\mathbf{S} \times \frac{\delta H}{\delta \mathbf{S}} - \epsilon \mathbf{S} \times \frac{d\mathbf{S}}{dt} \ . \tag{38}$$

However, we cannot simply add independent noise terms to these three equations, because they do not have the form of Langevin equations (all components of $d\mathbf{S}/dt$ appear in each equation due to the cross-product). Therefore we must first take all $d\mathbf{S}/dt$-terms to the l.h.s. before we can add three white-noise terms $\eta'_\alpha(\mathbf{r}, t)$:

$$\frac{d\mathbf{S}}{dt} = \frac{1}{1 + \epsilon^2 S^2} \left(-\mathbf{S} \times \frac{\delta H}{\delta \mathbf{S}} + \epsilon \mathbf{S} \times \left[\mathbf{S} \times \frac{\delta H}{\delta \mathbf{S}} \right] \right) + \eta' \tag{39}$$

with

$$\langle \eta'_\alpha(\mathbf{r}, t) \rangle = 0 \; , \tag{40}$$

$$\langle \eta'_\alpha(\mathbf{r}, t) \eta'_\beta(\mathbf{r}', t') \rangle = 2\epsilon k_B T \delta(\mathbf{r}' - \mathbf{r}) \delta(t' - t) \delta_{\alpha\beta} \; , \quad \alpha, \beta = 1, 2, 3 \; . \tag{41}$$

Now we take $\boldsymbol{\eta}'$ to the l.h.s. and undo the above procedure, i. e. we write (39) in the same form as (38), but with $\dot{\mathbf{S}} - \boldsymbol{\eta}'$ instead of $\dot{\mathbf{S}}$. Finally we get

$$\frac{d\mathbf{S}}{dt} = -\mathbf{S} \times \frac{\delta H}{\delta \mathbf{S}} - \epsilon \mathbf{S} \times \frac{d\mathbf{S}}{dt} + \boldsymbol{\eta} \tag{42}$$

with

$$\boldsymbol{\eta} = \boldsymbol{\eta}' - \epsilon(\mathbf{S} \times \boldsymbol{\eta}') \; . \tag{43}$$

Now we calculate the variances of $\boldsymbol{\eta}$ and see that the width of the distribution for the component parallel to the spin vector is $\sigma_0 = \sqrt{2\epsilon k_B T}$, while the widths for the perpendicular components are $\sigma_0 \sqrt{1 + \epsilon^2 S^2}$. Below we will apply the constraint $|\mathbf{S}| = 1$, which means here that only the perpendicular components are relevant. Thus we can replace $\boldsymbol{\eta}$ in (42) by $\boldsymbol{\eta}'$ if we correct the widths by a factor of $\sqrt{1 + \epsilon^2}$. But in our simulations we use values of ϵ in the order of 10^{-3}; therefore the correction factor can even be neglected in the following.

We form

$$\mathbf{S}\left(\frac{\partial \mathbf{S}}{\partial X_i} \times \boldsymbol{\eta}\right) = \left(\mathbf{S} \times \frac{\partial \mathbf{S}}{\partial X_i}\right)\boldsymbol{\eta} \; , \tag{44}$$

integrate over $\mathbf{r}$ and combine this with the results of the previous section, giving

$$\hat{\mathbf{A}}\ddot{\mathbf{X}} + \hat{\mathbf{M}}\ddot{\mathbf{X}} + \hat{\mathbf{G}}\dot{\mathbf{X}} = \mathbf{F} + \mathbf{F}^{st} \; . \tag{45}$$

For the stochastic force

$$F_i^{st} = \frac{1}{S^2} \int d^2 r \left(\mathbf{S} \times \frac{\partial \mathbf{S}}{\partial X_i}\right)\boldsymbol{\eta}(\mathbf{r}, t) \tag{46}$$

we need to calculate the mean $\langle F_i^{st} \rangle$ and the variance $\mathrm{Var}(F_i^{st})$. We define

$$F_i^{st} = \int d^2 r \, f_i^{(\alpha)} \eta_\alpha, \quad f_i^{(\alpha)} = \frac{1}{S^2} \epsilon_{\alpha\beta\gamma} S_\beta \frac{\partial S_\gamma}{\partial X_i} \tag{47}$$

with summations over repeated indices. The mean is zero and for the correlation functions[3] we get with (41)

$$\langle F_i^{st}(t) F_i^{st}(t') \rangle = 2\epsilon k_B T \delta(t - t') \int d^2 r f_i^{(\alpha)}(\mathbf{r}) f_i^{(\alpha)}(\mathbf{r}) \; . \tag{48}$$

Instead of the S_α we use the fields ϕ and ψ in (12) and thereby fulfill the constraint $|\mathbf{S}| = 1$. After some calculations we obtain

$$\mathrm{Var}(F_i^{st}) = 2\epsilon k_B T \int d^2 r \left\{ (1 - \psi^2)\left(\frac{\partial \phi}{\partial X_i}\right)^2 + \frac{1}{1 - \psi^2}\left(\frac{\partial \psi}{\partial X_i}\right)^2 \right\} \; . \tag{49}$$

---

[3] The correlation between different components of $\mathbf{F}^{st}$ can also be calculated and turns out to be zero.

Here the leading contribution comes from the static vortex structure given in (21):

$$\mathrm{Var}(F_i^{\mathrm{st}}) = 2\pi\epsilon k_{\mathrm{B}}T \int\limits_0^L \mathrm{d}r\; r \left\{ \frac{1 - \psi_0(r)^2}{r^2} + \frac{(\psi_0'(r))^2}{1 - \psi_0(r)^2} \right\} . \tag{50}$$

As ψ_0 decays exponentially, the second integral is independent of L, while the first one grows logarithmically. This would mean that the stochastic forces depend on the system size. We can avoid this by restricting the stochastic integral (46) to a circle with radius $\rho(T)$, i. e., we assume that for $r > \rho$ all contributions cancel because the thermal fluctuations are larger than the vortex structure. In fact, a look at the spin configuration gives this impression. For T of the order of the Kosterlitz-Thouless transition temperature T_{c} the vortices even appear to be structures with lengths of the order of a lattice constant [9]. Therefore we now tentatively replace the upper integration limit in (50) by the effective vortex radius $\rho(T)$. Moreover we divide the integral into an outer part from $a_c \le r \le \rho$ and a core part $C(a_c)$ and obtain

$$\mathrm{Var}(F_i^{\mathrm{st}}) = 2\epsilon k_{\mathrm{B}}T \cdot \pi \left\{ \ln\frac{\rho(T)}{a_c} + C(a_c) \right\} . \tag{51}$$

So far we have considered a vortex in the center of a large circular system. If the above assumption is correct, Eq. (51) also holds for an arbitrary position of the vortex center and for an arbitrary shape of the system (as long as the vortex center is not close to a boundary).

The results (48) and (51) then mean that the stochastic forces can be represented as white noise on the level of the collective coordinates with the properties $\langle F_i^{\mathrm{st}} \rangle = 0$ and

$$\langle F_i^{\mathrm{st}}(t) F_j^{\mathrm{st}}(t') \rangle = D_{\mathrm{eff}}\delta_{ij}\delta(t - t') \tag{52}$$

where the effective diffusion constant is identical to the r.h.s. of (51). The diffusion constant on the microscopic level is $D = 2\epsilon k_{\mathrm{B}}T$.

The effective vortex radius can be estimated in the following way: The vortex structure is destroyed where the thermal energy per unit area a_0^2 exceeds the vortex energy density E, i. e.

$$E(\rho) = \frac{f}{2}k_{\mathrm{B}}T/a_0^2 . \tag{53}$$

where $f = 2$ is the number of degrees of freedom per spin. E is obtained from the Hamiltonian density (18). We insert the static structure (21) and get an equation for $\rho(T)$ from (53)

$$\frac{1 - \psi_0(\rho)^2}{\rho^2} + \delta\left[4\psi_0(\rho)^2 - \psi_0'(\rho)^2 \right] + \frac{\psi_0'(\rho)^2}{1 - \psi_0(\rho)^2} = \frac{2k_{\mathrm{B}}T}{a_0^2 J S^2} . \tag{54}$$

For very low temperatures $(T \ll T_{\mathrm{c}})$ ρ is much larger than the core radius r_{v}, here we can neglect ψ_0 and obtain

$$\rho \approx a_0 \sqrt{\frac{JS^2}{2k_{\mathrm{B}}T}} \;.$$
(55)

For higher temperatures we evaluate (54) numerically using the following fit function which represents very well the static vortex structure in simulations at zero temperature [10]

$$\psi_0 = \frac{c_1 \mathrm{e}^{-z}}{\sqrt{z + c_2 \mathrm{e}^{-c_4 z} + c_3 \mathrm{e}^{-c_5 z^2}}}, \quad z = \frac{r}{r_{\mathrm{v}}}$$
(56)

with

$$c_1 = 2.17 \;, \quad c_3 = 1.33 \;, \quad c_5 = 0.27 \;,$$
(57)

$$c_2 = c_1^2 - c_3 \;, \quad c_4 = \frac{2c_1^2 + 1}{c_1^2 - c_3} \;.$$
(58)

We set a_0, J, S and k_{B} equal to one and plot $\rho(T)$ for different values of the anisotropy δ (Fig. 3). For very low temperatures $\rho(T)$ becomes independent of δ and follows Eq. (55). In Fig. 4 we plot the temperature for which ρ approaches one lattice constant. The dependence on δ is qualitatively the same as that of $T_{\mathrm{c}}(\delta)$, and the saturation value is quite close to the value $T_{\mathrm{c}} \approx 0.8$ for the XY-model. Both facts fit well to the above observation that near T_{c} the effective vortex radius seems to be in the order of the lattice constant.

Using the fitted ψ_0 from (56) and $\rho(T)$ as upper limit in (50) we can easily compute $D_{\mathrm{eff}} = \mathrm{Var}(F_i^{\mathrm{st}})$ as a function of T for various values of δ (Fig. 5). For low temperatures

$$D_{\mathrm{eff}} \approx \epsilon \, \alpha T^\gamma$$
(59)

is a good approximation where $\gamma \approx 0.8$ independent of δ and α decreases with δ. The exponent must be compared with $\gamma = 1$ in (50) if we integrate up to L.

4 Langevin Dynamics Simulations

We begin with one vortex on a circular shaped square lattice with a radius of L lattice points. We use free boundary conditions to get an image antivortex which leads to a small radial force on our vortex. The initial spin configuration stems from an iterative program which produces a moving vortex structure on the lattice [10]. In this way we avoid the radiation of spin waves which would appear during the first time units if we use a continuum approximation for the vortex shape.

For the time integration of the Landau-Lifshitz equation we use the discrete version of (39) where $\mathrm{d}\mathbf{S}/\mathrm{d}t$ has already been isolated on the l.h.s.. In contrast to our analytical calculations we work here with the cartesian components S_α. Therefore we explicitly take into account the constraint $\mathbf{S}^2 = 1$ by adding $\mathbf{S}$ times

158

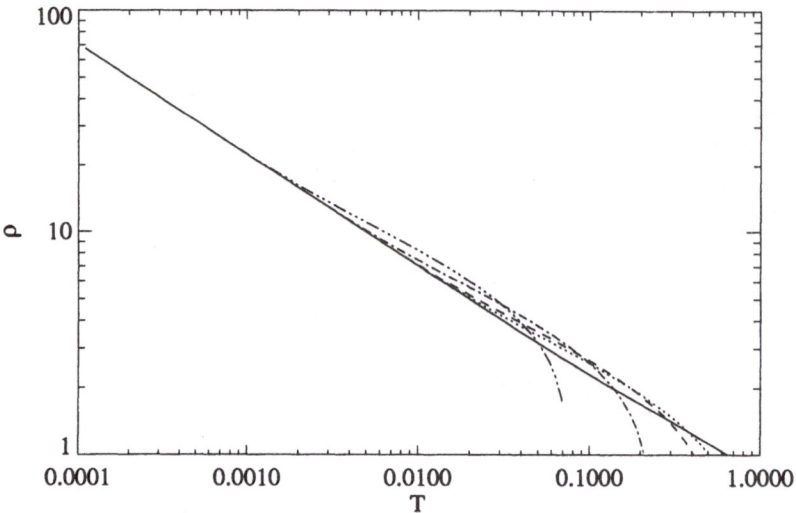

Fig. 3. Effective vortex radius ρ vs. temperature, ———: $\delta = 0.3$, $\cdots\cdots$: $\delta = 0.1$, $---$: $\delta = 0.07$, $-\cdot-\cdot-$: $\delta = 0.03$, $-\cdots-$: $\delta = 0.01$

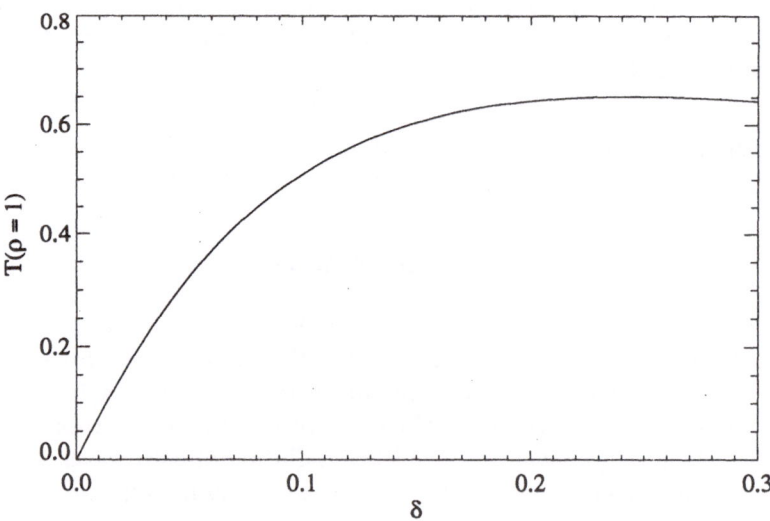

Fig. 4. The temperature for which the effective vortex radius approaches one lattice constant, vs. the anisotropy constant δ

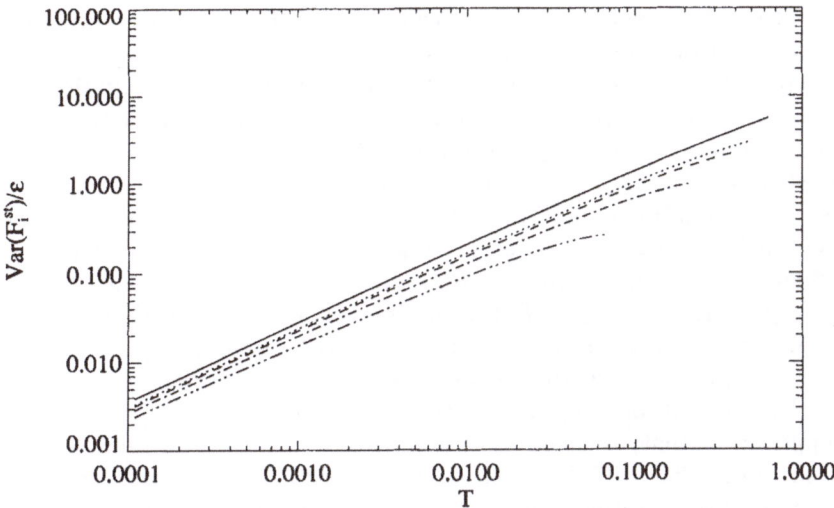

Fig. 5. Variance of stochastic forces over ϵ vs. temperature, ——: $\delta = 0.3$, ······: $\delta = 0.1$, — — —: $\delta = 0.07$, — · — · —: $\delta = 0.03$, — · · · —: $\delta = 0.01$

a Lagrange parameter λ to (39). We form the time derivative of the constraint, eliminate λ and get

$$\frac{\mathrm{d}}{\mathrm{d}t}\mathbf{S} = \mathbf{U} - \frac{\mathbf{SU}}{S^2}\mathbf{S} \tag{60}$$

with

$$\mathbf{U} = \frac{1}{1 + \epsilon^2 S^2}\left(-\mathbf{S} \times \frac{\delta H}{\delta \mathbf{S}} + \epsilon \mathbf{S} \times \left[\mathbf{S} \times \frac{\delta H}{\delta \mathbf{S}}\right]\right) + \boldsymbol{\eta}' , \tag{61}$$

where the site index has been omitted. We note that (60) is the same as orthogonalizing $\dot{\mathbf{S}}$ and $\mathbf{U}$ by the Gram-Schmidt method. For the time integrations we use the same code as in [1].

To find a proper damping constant we checked the time dependence of the system energy using different damping constants for $L = 24$ and $T = 0.02$. The energy at $t = 0$ is the same as for $T = 0$ and $\epsilon = 0$ because the noise will be introduced with the first time step of the simulation. The energy then rises and saturates on a value independent of ϵ, but for $\epsilon > 8 \cdot 10^{-3}$ the energy decreases slowly after saturation. The saturation time gets longer with lower ϵ, for $\epsilon \geq 2 \cdot 10^{-3}$ we get acceptable saturation times < 300 (in units of $\hbar/(JS)$).

The difference between the energy without temperature and the saturation energy with temperature must be the thermal energy. We computed the mean thermal energy per spin at several temperatures and it agreed with $f/2 \cdot k_{\mathrm{B}}T$ up to $T = 0.9$. For higher temperatures we get too low values for the energy. We believe that the numerical procedure would have to be improved if we were interested in this regime.

We studied the trajectory of the vortex center at different temperatures keeping $L = 24$ and $\epsilon = 3 \cdot 10^{-3}$ fixed. We can distinguish three temperature regimes in which the trajectories differ qualitatively:

For $0 \leq T < T_3 \approx 0.05$ we observe two frequencies in the oscillations around the mean trajectory which can be identified with the cycloidal frequencies $\omega_{1,2}$ in (34). The intensities of $\omega_{1,2}$ decrease with temperature and vanish at T_3 in the background, but $\omega_{1,2}$ are constant in the whole regime. This means that here the 3rd-order equation of motion (45) with temperature-independent parameters can describe the vortex dynamics. For one temperature of this regime we plot in Fig. 6 and 7 the radial coordinate $R(t)$ and the azimuthal displacement $\varphi(t) = \phi(t) - \omega_0 t$. Here $\omega_0 = F_0/(GR_0)$ is the frequency of the rotation on the mean trajectory which is essentially a circle where the radius R_0 grows very slowly with rate gF_0/G^2 due to the damping. On the mean trajectory the vortex is driven by a radial force F_0 due to the image vortex at $R^{(i)} = L^2/R_0$, which has opposite vorticity but the same polarization [11]. As the average motion is very slow ($\omega_0 \approx 2.5 \cdot 10^{-3}$) we can actually work in a cartesian system and use the results (30)-(34). Here the X_1-axis points in the radial direction, and the X_2-axis in the azimuthal direction. Fig. 8 shows the Fourier spectrum of $R(t)$. In addition to $\omega_{1,2}$ one also observes the difference $\Delta\omega = \omega_2 - \omega_1$. This can be explained by solving the 3rd-order equation of motion in polar coordinates, which is not shown here because the formulas are too cumbersome.

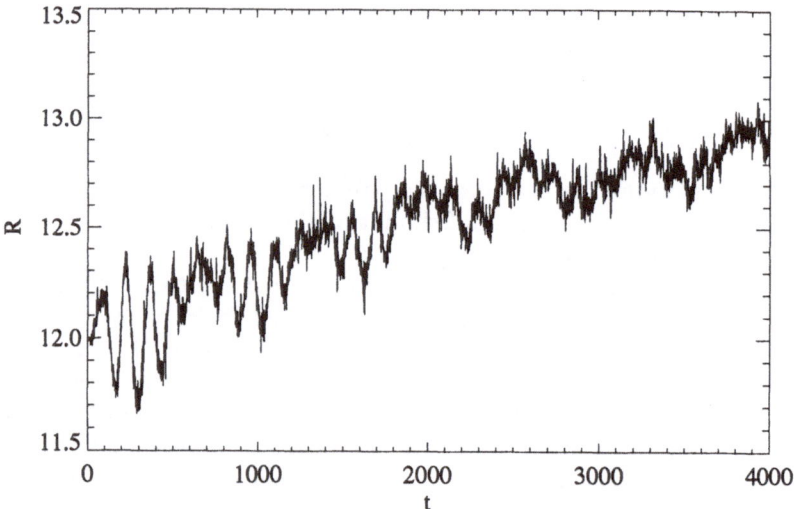

Fig. 6. Radial coordinate vs. time in units of $\hbar/(JS)$, for $L = 24$, $T = 0.002$ and $\epsilon = 0.003$

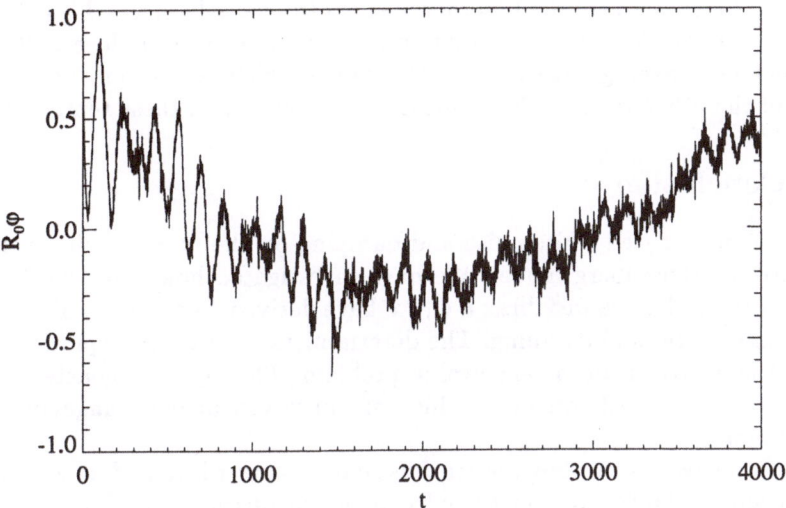

Fig. 7. Displacement in azimuthal direction vs. time, for the same vortex as in Fig. 6

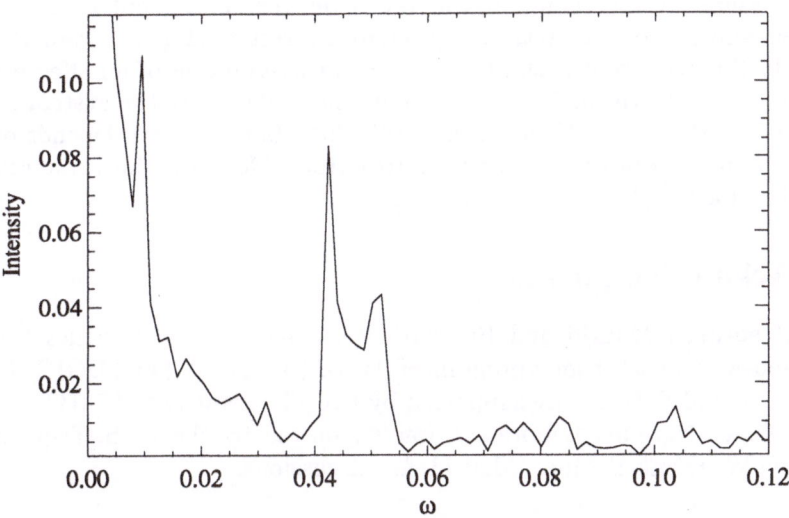

Fig. 8. Fourier spectrum of the radial coordinate, from the simulation data also used for Fig. 6 and Fig. 7 $(0 < t < 5000)$

For intermediate temperatures $T_3 < T < T_1 \approx 0.1$ no cycloidal frequencies are observed; the vortex motion is here identical to the average motion discussed above. This means that the 1st-order part of (45), namely $\hat{\mathbf{G}}\dot{\mathbf{X}} = \mathbf{F}+\mathbf{F}^{st}$ describes the motion. For $T > T_1$ the noise is very large, but at least close to T_1 one can still define an average trajectory of the vortex which seems to move in the mean field of the other vortices that appear spontaneously at these temperatures.

5 Conclusion

We have introduced white noise and damping on the microscopic level of a 2D anisotropic Heisenberg model. We have investigated how this transfers to the level of the collective variables; i. e., we have derived an equation of motion with stochastic forces and damping. The deterministic part of this equation has been solved and gives a linear eigenvalue problem. The eigenfrequencies have been identified with the frequencies which are observed in our Langevin dynamics simulations.

As we have shown that the stochastic forces are white and as we know their mean and variance, we can actually go much further now: We can solve the stochastic equation of motion (45) (It is equivalent to a system of 6 coupled linear Langevin equations). This means that we can calculate analytically the variances of the collective variables

$$\text{Var}(X_i) = \langle X_i(t)^2 \rangle - \langle X_i(t) \rangle^2 \ , \tag{62}$$

which will depend on the effective diffusion constant D_{eff}. On the other hand we can measure these variances in our Langevin dynamics simulations. The comparison will test our analytical and numerical results (Eq. (51) and Fig. 5) for D_{eff}. In this way we can clarify whether our assumption of an effective vortex radius $\rho(T)$ is justified. If not, $\rho(T)$ must be replaced by the system radius L. In this case the stochastic forces are still white, but $\text{Var}(F_i^{st})$ depends on L and on the radius R_0 of the mean vortex trajectory. Moreover the variances for the radial and azimuthal directions are different.

6 Acknowledgments

Travel between Madrid and Bayreuth was financed by "Acciones Integradas Hispano-Alemanas", a joint program of DAAD (Az. 314-AI) and DGICyT (HA96-9B). A. S. and F. D.-A. are supported by CICyT (grant MAT95-0325). Work at Los Alamos National Laboratory was supported by the U. S. Department of Energy. We thank E. Moro, Madrid, for discussions.

References

1. F. G. Mertens, H.-J. Schnitzer and A. R. Bishop, *Hierarchy of equations of motion for nonlinear coherent excitations, applied to magnetic vortices*, submitted to Phys. Rev. B (1996)

2. A. A. Thiele, *Steady State Motion of Magnetic Domains*, Phys. Rev. Lett. **30**, 230-233 (1973)

3. A. A. Thiele, *Applications of the Gyrocoupling Vector and Dissipation Dyadic in the Dynamics of Magnetic Domains*, J. Appl. Phys. **45**, 377-393 (1974)

4. D. L. Huber, *Dynamics of Spin Vortices in Two-Dimensional Planar Magnets*, Phys. Rev. **B26**, 3758-3765 (1982)

5. S. Iida, *The difference between Gilbert's and Landau-Lifschitz's equations*, J. Phys. Chem. Solids **24**, 625-630 (1963)

6. G. M. Wysin, F. G. Mertens, A. R. Völkel and A. R. Bishop, *Mass and Momentum for vortices in two-dimensional easy-plane magnets*, in *Nonlinear coherent structures in physics and biology*, ed. by K. H. Spatschek and F. G. Mertens (1994)

7. A. R. Völkel, F. G. Mertens, A. R. Bishop and G. M. Wysin, *Motion of vortex pairs in the ferromagnetic and antiferromagnetic anisotropic Heisenberg model*, Phys. Rev. **B43**, 5992-6005 (1991)

8. M. E. Gouvea, G. M. Wysin, A. R. Bishop and F. G. Mertens, *Vortices in the two-dimensional anisotropic Heisenberg model*, Phys. Rev. **B39**, 11840-11849 (1989)

9. S. Miyashita, H. Nishimori, A. Kuroda and M. Suzuki, *Monte Carlo Simulation and Static and Dynamic Critical Behavior of the Plane Rotator Model*, Progress of Theor. Phys. **60**, 1669-1685, (1978)

10. H.-J. Schnitzer, *Zur Dynamik kollektiver Anregungen in Hamiltonschen Systemen*, Ph.D.-thesis, University of Bayreuth (1996)

11. F. G. Mertens, G. M. Wysin, A. R. Völkel and A. R. Bishop, *Cyclotron-Like Oscillations and Boundary Effects in the 2-Vortex Dynamics of Easy-Plane Magnets*, in *Nonlinear coherent structures in physics and biology*, ed. by K. H. Spatschek and F. G. Mertens (1994)

Chaotic Dynamics
in Classical Lattice Field Theories

T.S. Biró[1], Á. Fülöp[1], C.Gong[2], S. Matinyan[2], B. Müller[2] and A. Trayanov[2]

[1] MTA KFKI RMKI Budapest, Hungary
[2] Duke University, Durham, NC, USA

Abstract. Classical equations of motion of many field theories (compact U(1), SU(2) and SU(3) Yang-Mills, SU(2) Yang-Mills-Higgs, O(4)) show a chaotic dynamical behavior of homogeneous time-dependent configurations. The Hamiltonian lattice regularization of such theories were studied by us numerically and the leading Lyapunov exponent – a real-time property – has been extracted. Its scaling with the average energy per plaquette (interpreted as temperature) permits to draw conclusions in the fine resolution (vanishing lattice spacing) limit. We also report about a coincidence between the leading Lyapunov exponent in SU(2) Yang-Mills lattice theory and the resummed perturbative gluon damping rate at high temperature ($\mathcal{O}(g^2T)$).

1   Introduction

Unlike the most contributions to this workshop this paper deals with highly excited states of lattice field models. At high temperatures, simulated by high energy-density in the system, the thermal distribution of bosonic quanta shows an enhancement of the low momentum modes:

$$n(\omega) = \frac{1}{e^{\omega/T} - 1} \approx \frac{T}{\omega} \quad . \tag{1}$$

This hints at the idea that such a system can be close to the classical limit. On the other hand the study of real-time dynamics of fields is not yet in the stage of doing numerical calculations on the full quantum mechanical system. Instead, the thermal equilibrium state is studied. As a result of these circumstances a numerical and analytical study of *classical* lattice gauge field theory has been developed in recent years, see (Biró, Matynian and Müller 1994) for details.

The issue of the investigation here is the real-time dynamics of such classical Hamiltonian systems, in particular their evolution from an ordered state towards a thermalized one. During this process entropy is produced, which in principle can be observed as enhanced produced particle abundances in accelerator experiments or may have consequences for cosmological models.

This paper is organized as follows: first, basic concepts of chaotic dynamics and their application in simple few-mode models of nonabelian interactions are reviewed. Next, classical Hamiltonian lattice field theory is discussed: definitions, constraints and algorithmical highlights. Finally, results on the chaotic divergence of initially close field-configurations on the lattice and its dependence on physical circumstances, like the energy, are presented.

2   Simple Chaotic Systems

Chaotic behavior is generally identified using the "small-action large-effect" principle: a tiny deviation in the initial state leads to an exponentially diverging position in the phase space in the course of time evolution. Each Hamiltonian system can be characterized by a set of conjugate variables, they span out the phase space. In this high dimensional space (for conservative systems always even) each point describes a possible state of motion. The solutions of the equations of motion can be viewed as graphs of trajectories in this space.

The most characteristic quantity that measures chaos is the Lyapunov exponent. It describes the rate of exponential divergence of initially adjacent points at a distance of $d(0)$ at time 0 and of $d(t)$ at time t:

$$\lambda = \lim_{t \to \infty} \lim_{d(0) \to 0} \frac{1}{t} \ln \frac{d(t)}{d(0)} \quad . \tag{2}$$

For a conservative system, as all Hamiltonian lattice field systems up to now investigated are, the sum of all Lyapunov exponents must be zero due to Liouville's theorem:

$$\sum \lambda = 0 \quad . \tag{3}$$

Nevertheless, assuming a finite resolution of the phase space (which may eventually stem from quantum mechanics) the negative Lyapunov exponents, which describe a shrinking of an initial area of close points in the phase space, cannot be seen by an observer. This loss of information is exactly compensated in the conservative case by the inflation of the phase space area in other directions. Therefore, the natural measure of the information loss, or entropy generation, is the sum of all positive Lyapunov exponents

$$\sigma_{\mathrm{KS}} = \sum_{\lambda > 0} \lambda \quad . \tag{4}$$

Quantitiy (4) is called the Kolmogorov-Sinai entropy.

This correspondence between microscopic information loss due to the relative motion of close trajectories and the physical entropy generation rate allows us to relate the study of classical dynamics on lattices to high-temperature field theory. The correspondence between energy E and the temperature T in a final state with maximized entropy on the one hand and the correspondence between the entropy-density generation rate $\dot{s}$ and the gross value of the Lyapunov exponents σ_{KS} on the other hand tempts us to relate the universal results of pure classical simulations to physical parameters of a hot field system.

After this brief review of some basic concepts of chaotic dynamics let us demonstrate their use in simple examples. The simplest analytic model, that shows reminiscent properties of Yang-Mills theory, is the xy-model. Considering only two homogeneous modes of the nonabelian SU(2) vector potential, namely

$x = A_1^2(t)$ and $y = A_2^1(t)$, as nonzero we arrive at the following simple effective Hamiltonian density

$$H = \frac{1}{2}\left(\dot{x}^2 + \dot{y}^2\right) + \frac{g^2}{2}x^2y^2 \quad .$$

(5)

Multiplication by g^2 and scaling by the coupling g,

$$gx \to x \quad , \quad gy \to y \quad , \quad g^2H \to H \quad ,$$

(6)

leads to the universal Hamiltonian. Any classical motion with fixed energy, e.g., $H = 1/2$, has its turning points on the hyperbolas satisfying $x^2y^2 = 1$.

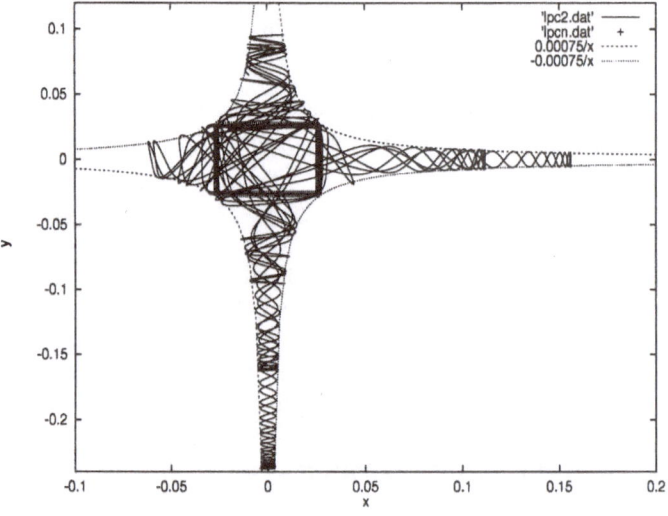

Fig. 1. A classical trajectory in the xy-model.

It is simple to understand the chaoticity of this system by inspecting the negative (inward) curvature of the wall of turning points. Such a concave shape defocuses the nearby incoming trajectories. On the other hand, this shape corresponds to negative eigenvalues of the stability matrix $\partial^2 V/\partial q_i \partial q_j$.

Explicit numerical simulations of the motion of several hundred points has also revealed the scattering over the whole available phase space after a transition period. As a slight modification of this Hamiltonian, in analogy to the coupling of a Higgs field to a Yang-Mills field, a mass term may be introduced. This has a regularizing effect. In fact, an analog of the Reynolds number,

$$R = \frac{1}{4}\frac{g^4v^4}{g^2E/V} \quad ,$$

(7)

could be employed. Above a given value of the control parameter (7) no more chaotic motion occurs in the xy-model and in analogous few-mode models.

These investigations have recently been extended to few-mode models which simulate the interaction of short- and long-wavelength plasma excitations. The original Yang-Mills Hamiltonian,

$$H = \int d^3x \frac{1}{2} \left(E_{ia} E_{ia} + B_{ia} B_{ia} \right) \quad , \tag{8}$$

was studied with an ansatz that coupled three $k = 0$ soft modes to six $\mathbf{k} = (0, 0, T)$ hard modes. Hard modes are usually considered as having a momentum and an energy in the order of the temperature T (Blaizot and Iancu 1994). Such an ansatz is described in detail in Ref.(Biró, Matinyan and Müller 1995).

The Hamiltonian ansatz consists of three physically different parts: a part describing hard modes alone, like oscillators with frequency $\omega = T$, a soft part resembling the xyz-model (3 degrees of freedom version of the xy-model described above) and finally the interaction part. The latter is responsible for effective masses of the soft x, y and z modes. Moreover, an intrinsic instability driven by the interaction with the hot part arises, occurring as an xy off-diagonal mass term,

$$\frac{1}{V} H_{\text{soft,hard}} = \frac{1}{2} \left(\dot{x}^2 + \dot{y}^2 \right) + \frac{m^2}{2} \left(x^2 + y^2 + 2cxy \right) \quad . \tag{9}$$

Here the magnitude of the off-diagonal mass term can exceed that of the diagonal ones ($|c| > 1$): a genuine sign for instability.

Whether the thermally generated mass - as reflected in the one-loop resummed perturbation theory at high temperature - or the instability due to the stochastic interaction with members of a heat bath dominates the physical behavior of the few slow modes under inspection, is a question to the full theory. On the basis of the semi-simple example discussed here we can distinguish the following three physical time scales:

i) the hard modes oscillate fast, the characteristic time is in the order of $t_1 \sim 1/T$,

ii) the soft modes evolve on a time scale according to their generated thermal mass, $t_2 \sim 1/gT$,

iii) while finally the stochasticity of the interaction with the heat bath, i.e. the random phases of the hard oscillators when interacting with them, averages out in an even longer time scale $t_3 \sim 1/g^2T$.

The time scale t_3 corresponds to the chaotic evolution of the classical subsystem of slow modes.

Recent investigations in thermal field theory indicate that the above experience is more general. Let us recapitulate briefly the results of Greiner and Müller(Greiner and Müller 1996) on the scalar Φ^4 theory.

In the effective description of $k = 0$ modes – besides the classical equation of motion directly derivable from the bare Lagrangian – different terms occur which are perturbatively calculable order by order in the high-temperature loop expansion. Besides the already well known generation of in-medium mass and plasmon damping in the Markovian limit stochastic forces can be derived from the two-loop self-energy. More precisely, in the classical limit the damping and the strength of the stochastic source term are connected by the general fluctuation-dissipation theorem, reducible to the Kubo formula for low frequency soft modes.

These stochastic forces occurring (i) in the perturbative field theory at high temperature when inspecting the closely classical soft sub-sector as well as (ii) in non-perturbative few-mode models of chaotic dynamics in nonabelian field systems motivate us strongly to investigate the classical real-time dynamics of extended field systems by non-perturbative methods like the lattice regularization. Its basic concepts and methods are reviewed in the next section.

3 Classical Lattice Gauge Theory

Classical Lattice Gauge Theory (CLGT) (Biró et.al. 1994), (Biró, Gong and Müller 1995a) works with the same variables as its quantum equilibrium version. The difference is in the description of the dynamics: here, real-time differential equations are solved for the basic variables and these solutions are studied as functions of constants of motion, most often the total energy. No Monte Carlo sampling is done, rather the lattice dynamics itself drives the system towards higher entropy states. This real-time dynamics finds the lattice system most of the time in probable states; they are, however, not weighted by the Euclidean action but governed by the Minkowskian energy.

In a three-dimensional regular lattice with lattice spacing a the basic variables are group elements related to the vector potential

$$U_{x,i} = \exp\left(a A_i^c(x) T^c\right) \quad , \tag{10}$$

with T^c being the group generators. For SU(2) e.g., they are given by the Pauli matrices τ: $T^c = -(ig/2)\tau^c$. The indices x, i denote a link of the lattice staring at the three dimensional position (site) x and pointing into the i-th direction. The product of such group elements along any connected line is a gauge covariant quantity; the traces over such products along a closed path are invariant. This lattice system is therefore very suitable for describing gauge theories.

The nonabelian magnetic field strength is related to the plaquette product,

$$U_{x,ij} = U_{x,j} U_{x+j,i} U_{x+i,j}^\dagger U_{x,i}^\dagger \quad , \tag{11}$$

in the familiar way

$$U_{x,ij} = \exp\left(\epsilon_{ijk} a^2 B_k^c T^c\right) \quad . \tag{12}$$

The electric energy in Hamiltonian lattices can be defined in two ways (left-handed and right-handed in the group),

$$E_{x,i}^c = \frac{2}{ag^2} \mathrm{tr}\left(T^c \dot{U}_{x,i} U_{x,i}^\dagger\right) \quad ,$$

$$E_{x,i}^c = \frac{2}{ag^2} \mathrm{tr}\left(T^c U_{x,i}^\dagger \dot{U}_{x,i}\right) \quad . \tag{13}$$

In the continuum limit the difference should not matter, however. In real-time simulations we use the canonical momentum variables on each link,

$$P_{x,i} = \dot{U}_{x,i} \quad .$$ (14)

The lattice Hamiltonian can be written as a sum over single-link contributions,

$$H = \sum_{\text{links}} \left[\frac{1}{2}\langle P, P \rangle + \left(1 - \frac{1}{4}\langle U, V \rangle \right) \right] \quad ,$$ (15)

where $\langle A, B \rangle = \text{tr}(AB^\dagger)$ denotes the scalar product in the group space, and $V_{x,i}$, the complement link variable, is constructed of products of U-s along all link triples which close with the given link (x, i) an elementary plaquette. In three dimensions each link has four attached plaquettes, thus in the link sum every plaquette would be counted four times. This is compensated by the factor $1/4$ in Eq. (15).

Besides the Hamiltonian (15) given above the time evolution must satisfy several constraints. Two of them, namely the unitarity of SU(N) group elements, $\langle U, U \rangle = 1$ and their conservation, $\langle U, P \rangle = 0$, are ensured by introducing Lagrange multipliers. A nontrivial constraint is Gauss' law,

$$\Gamma = \sum_+ PU^\dagger - \sum_- U^\dagger P = 0 \quad ,$$ (16)

where the first sum (with $+$) is understood to be over links starting at the site where Γ belongs to and the second sum (with $-$) corresponds to links ending at that site. This quantity is automatically conserved by the Hamiltonian equations of motion on the lattice,

$$\dot{\Gamma} = \sum_+ VU^\dagger - \sum_- U^\dagger V = 0 \quad ,$$ (17)

its initially zero value is the only nontrivial point to install. The Figure below shows a typical random initial condition of an SU(2) lattice theory. The grayness of the links are proportional to the energy corresponding to that link.

The computational algorithm that solves the Hamiltonian equations of motion works, of course, with discrete time steps. It is possible to identify certain classes of update algorithms which satisfy Noether's theorem also for discrete time steps and hence to choose an algorithm which satisfies Gauss law exactly, i.e., up to the greatest precision representable on the computer irrespective of the algorithmical time step dt. Details of this derivation are given elsewhere (Biró 1995).

The update of link variables has been derived from the following implicit recursion relations

$$U' = U + (P' - \epsilon U) \quad ,$$
$$P' = P + (V - \mu U + \epsilon P') \quad .$$ (18)

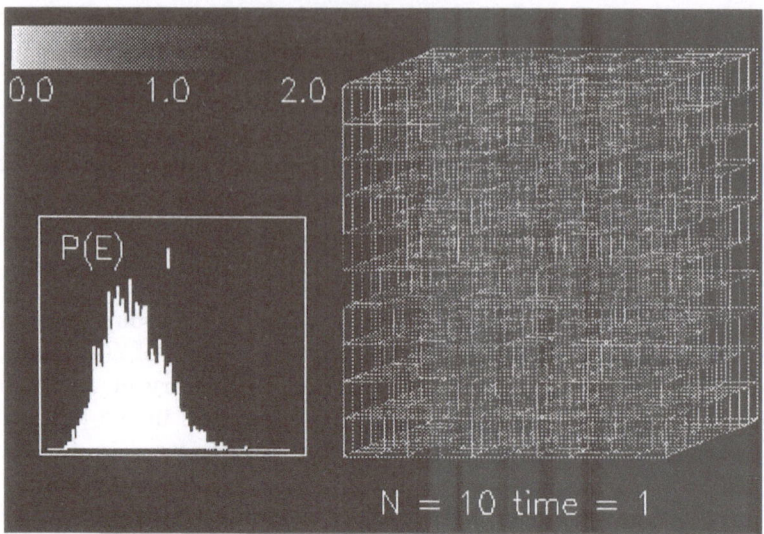

Fig. 2. Energy distribution in a state uniformly random in the SU(2) group space. The darkness of the links is proportional to the energy density.

Here, the primed values are the updated ones and the Lagrange multipliers are given by

$$\epsilon = \langle U, P' \rangle \quad , \quad \mu = \langle U, V \rangle + \langle P', P' \rangle \quad . \tag{19}$$

This implicit algorithm is analytically resolvable into the following line of explicit steps:

```
(V -= <U,V> U; P += V; c = sqrt{1+<P,P>}; P += U;
P /= c; U *= c; P -= U; U += P; in C language notations.)
```

In earlier calculations also other algorithms have been used, e.g. a Runge-Kutta algorithm of fourth order without special treatment of Gauss' law. The cumulated numerical violations of Gauss' law and of the energy conservation were found to be tolerable throughout.

With the above tool for initializing and investigating the time evolution of classical lattice gauge field systems, only the definition of a distance measure is left for studying chaotic properties. We use the gauge invariant definition based on the local magnetic energy difference:

$$D(U, U') = \frac{1}{2N_P} \sum_{x,ij} \left| \mathrm{tr} U'_{x,ij} - \mathrm{tr} U_{x,ij} \right| \quad . \tag{20}$$

Other, topologically equivalent distance measures are also imaginable. Their separation rate in case of chaotic behavior should not depend on details of the definition.

4 Results

Randomly chosen initial configurations on any finite lattice have a maximal distance because of the compactness of the plaquette variable $U_{x,ij}$. This is in contrast to the continuum theory, where in principle arbitrary large magnetic fields may occur. The system size also determines the distribution of the distance between two equally randomly chosen initial configurations. Restricting our recent analysis to purely magnetic configurations, we realize that a lattice of linear size of $N = 20$ is already large enough for having almost each pair of configurations at large distance. After ensuring this we compare the time evolution of a randomly chosen configuration with a nearby one.

The general pattern of the evolution of the logarithmic distance between two initially adjacent configurations is as follows: after a short period of irregular oscillations an exponential divergence sets in, finally a saturation of the distance on the compact energy surface limits this behavior. ¿From the linear part of the $logD$ - t/a curves we can read off the leading (maximal) Lyapunov exponent. A collection of these values at different energies at which the system was initialized reveals the scaling behavior of the chaotic dynamics. Fig.3 shows a recent collection of data on an SU(2) system going to smaller energies and using more precise numerical methods (rescaling) for measuring the leading Lyapunov exponent than before.

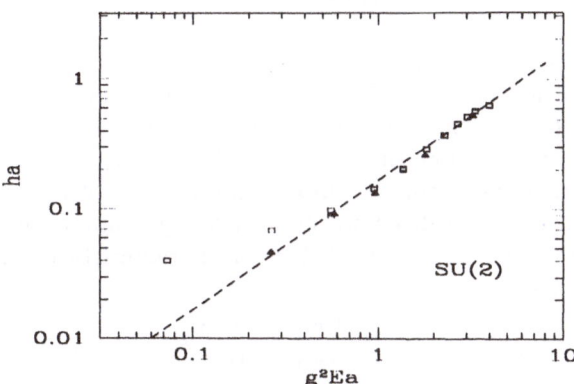

Fig. 3. The dependence of the scaled leading Lyapunov exponent on the scaled classical energy shows universal behavior.

The linear correspondence shown in figure 3 allows us to conjecture that the scaled quantities are proportional

$$a\lambda_0 \approx \frac{1}{6}g^2 Ea \quad . \tag{21}$$

In the fine resolution limit, $a \to 0$ their ratio is constant over a wide range of classical energy $g^2 E$. We believe to measure in this limit the leading Lyapunov exponent of lattice gauge field theories, in units of the physical energy $g^2 E$.

This linear correspondence is generic for nonabelian theories. For SU(3) Yang-Mills systems we obtain

$$\lambda_0 \approx \frac{1}{10}g^2 E \quad . \tag{22}$$

For relatively small lattices ($N = 2$ and $N = 3$) we were able to obtain the full Lyapunov spectrum of 144 and 256 modes, respectively. The typical structure of such a spectrum distinguishes three areas: about equally many modes with positive, zero, and negative Lyapunov exponents. The zero Lyapunov exponents reflect the gauge symmetry present in the dynamics; one third of the naively possible degrees of freedom are symmetries (Killing vectors). The positive-negative symmetry of the spectrum is due to the conservativeness of the Hamiltonian system.

Given the whole Lyapunov spectrum the Kolmogorov - Sinai entropy can be obtained. We arrive at the entropy production-rate,

$$\dot{s} = \frac{1}{N^3} \sum_i \lambda_i \approx 2\lambda_0 \quad . \tag{23}$$

A study of the plaquette energy-distribution revealed that the final state of the chaotic evolution of the SU(2) Yang-Mills lattice system is thermal with a distribution $P(E) \propto E^3 e^{-E/T}$. In the state after the initial divergence of trajectories, i.e. after a time of the order of $1/\lambda_0$ the magnetic and electric energy were approximately equal, equipartition was established. Comparing the spectral temperature of this distribution, the leading Lyapunov exponent and the entropy-generation rate we concluded that the characteristic entropy-production time can be given as

$$1/\tau_{KS} = \frac{\dot{\sigma}_{KS}}{\sigma_{KS}} \approx \frac{1}{12}g^2 T \quad . \tag{24}$$

Obtaining the dressed coupling g from one-loop perturbative renormalization we arrived at times of fractions of a fermi (0.2 - 0.4 fm/c) at several hundred MeV temperatures characteristic for heavy-ion collisions at present and future accelerators.

It is also interesting to note that the maximal Lyapunov exponent when related to the final temperature turned to be equal to the gluon damping rate[1].

---

[1] When defined properly, as the damping of energy carried by the plasmon(Matinyan and Müller 1996)

One quantity (the Lyapunov exponent) was obtained in purely classical, however non-perturbative calculation, while the other (the plasmon damping rate at $k = 0$) in an analytic resummed perturbative calculation. In due of time we have attempted several explanations of this coincidence (Biró, Gong and Müller 1995b) , but a deeper understanding of the correspondence between time forward classical initial value calculations and the hard thermal-loop expansion of self-energy calculations are still awaiting further investigations.

Finally, for the sake of comparison let us present briefly a few results on abelian and non-gauge field lattice systems (Biró et.al. 1994). In the case of compact U(1) lattice theory, there is chaotic behavior in the strong coupling phase, but it vanishes at small scaled energies $\propto a^2$ with the lattice spacing in the weak-coupling phase. The continuum U(1) theory is just abelian QED, independent oscillators are not chaotic. Coupling to the Higgs fields (Biró and Thoma 1996) and (Heinz et.al. 1996), important in standard model studies, does not change essentially the chaoticity of the Yang-Mills sector; the generated mass is negligible in the high-temperature phase and the stochastic forces become important. It is not trivial to extract the Higgs Lyapunov exponents and relate them to the Higgs damping rate because the interaction mixes the modes. An on-line diagonalization procedure of the monodromy matrix is unfortunately undoable with recent computational facilities.

Finally, on the way towards including matter fields, especially fermions in these investigations, we studied the effect of static charges on the chaotic dynamics of lattice Yang-Mills systems (Biró 1996), (Avakyan et.al. 1983) and (Lavkin 1996). In order to satisfy the non-abelian Gauss law with a prepared configuration of U-fields on the links one first defines a flux line along which Gauss's law can be recursively satisfied. Let us denote this chain of lattice sites by $(x_1, x_2, \ldots x_M)$. Imagine we have a charge Q at the first position. On the first link starting at this position the conjugate momentum field has to be

$$P_1 = Q \cdot U_1 \ . \tag{25}$$

Recursively, on any new link along the flux line, Gauss' law requires

$$P_n = U_{n-1}^\dagger \cdot P_{n-1} \cdot U_n \ . \tag{26}$$

At the end of the flux line we arrive at a closing charge of

$$Q' = -F^\dagger \cdot Q \cdot F \ , \tag{27}$$

where F is a product of group elements along the flux line (a Wilson line),

$$F = \prod_n U_n \ . \tag{28}$$

In order to have a color-neutral state, as all physical states are, one requires

$$Q + Q' = 0 \ . \tag{29}$$

Together with the constraint $\text{tr}Q = 0$ ensuring that Q describes a *color* charge with no neutral component, we finally obtain,

$$Q = \frac{q}{2}\left(F^{\dagger} - F\right) \quad .$$ (30)

So the static charge we can put on the lattice is mainly defined by the Wilson line, only its strength is an adjustable parameter.

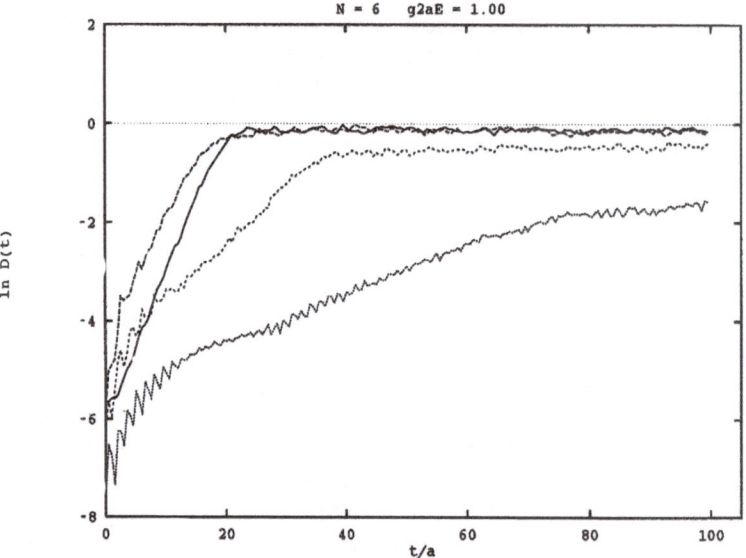

Fig. 4. The divergence of the logarithmic distance of initially random SU(2) field configurations in the presence of weak and strong static charges on the edges of the lattice diagonal.

This static charge shows a slow growing, but over a critical value dramatic effects on the chaotic dynamics of the Yang-Mills system appear. Not only the divergence of initially close configurations is slowed down, but also fluctuations tend to grow with increasing strength of the Wilson line. This strong presence of the static electric (Coulomb) sector also changes the equipartition between electric and magnetic energy: a large non-dynamical part does not equilibrate. Finally it is interesting to note that the change of the maximal Lyapunov exponent as a function of the charge of the Wilson line shows a characteristic S-shape known from first-order phase transitions (Fig.5).

5   Conclusion

We presented basic concepts of classical Hamiltonian lattice dynamics with special emphasis on the chaotic behavior and its possible survival in the continuum

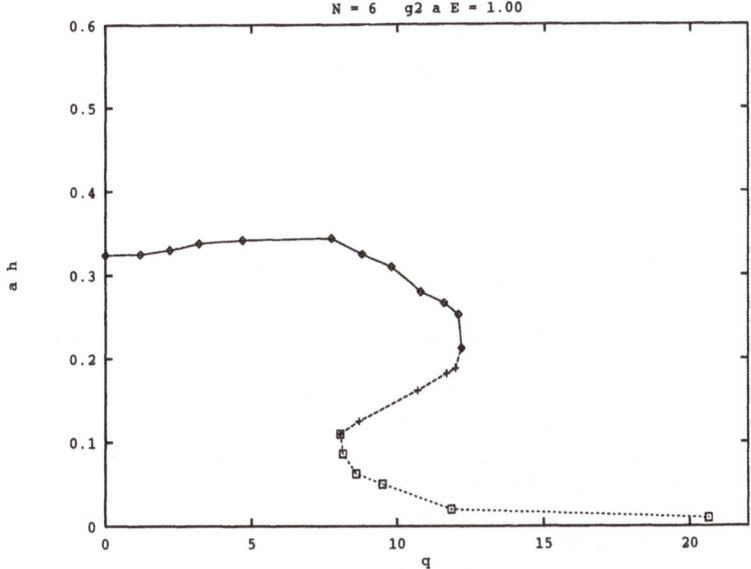

Fig. 5. The scaled maximal Lyapunov exponent of the lattice Yang-Mills system as a function of the charge of a cross-diagonal Wilson line.

limit. Studies of few-mode models helped us to distinguish three different time scales in coupled non-perturbative soft systems and in thermalized hard systems, which presumably correspond to different levels in a resummed loop expansion of hot field theory.

A numerical investigation of classical lattice Hamiltonian systems found that several field theory models can be chaotic. The nonzero maximal Lyapunov exponent scales with lattice spacing and the total energy behaves linearly in case of non-abelian gauge theories. This let us surmise that chaoticity is also present in the continuum limit. Some other theories, such the U(1) model, are chaotic on the lattice but not in this limit.

In simulations of SU(2) and SU(3) Yang-Mills systems we found numerical coincidence between the maximal Lyapunov exponent and the gluon damping rate obtained in resummed perturbative field theory. It supports the idea that the hard thermal loop-expansion describes semiclassical transport coefficients, which can be obtained from a classical dynamical simulation at high energy-density as well.

As a general feature chaotic dynamics provides a natural mechanism for entropy generation due to soft field dynamics: a promising perspective for the investigation of collective states of a quark-gluon plasma in relativistic heavy-ion collisions. In particular recent investigations on a possible disoriented chiral condensate formation in a heavy ion collision may relate the chaotic dynamics of the soft sub-system in the O(4) model with properties of the heat bath of hard pions present. The entropy generation by heating the soft collective modes in this case is escorted by particle production out of them.

References

A. Avakyan, S. Arutyunyan and G. Baseyan: *Elimination of Chaos of the Classical YM theory by external static charge*, Yerevan, EFI-641/31/-83, 1983

T.S. Biró, S.G. Matinyan and B. Müller: *Chaos and Gauge Field Theory*, World Scientific 1994

T.S. Biró, C. Gong, B. Müller and A. Trayanov: *Real-time Dynamics of Yang-Mills Theories on a Lattice*, Int. J. Mod. Phys. C **5**, 113, 1994

T.S. Biró, C. Gong and B. Müller: *Variational approach to real-time evolution of Yang-Mills gauge fields on the lattice*, Nucl. Phys. A **581**, 598, 1995

T.S. Biró, C. Gong and B. Müller: *Lyapunov Exponent and Plasmon Damping Rate in Nonabelian Gauge Theories*, Phys. Rev. D **52**, 1260, 1995

T.S. Biró: *Conserving Algorithms for Real-Time Nonabelian Lattice Gauge Theories*, Int. J. Mod. Phys. C **6**, 327, 1995

T.S. Biró, S.G. Matinyan and B. Müller: *Chaos driven by soft-hard mode coupling in thermal Yang-Mills theory*, Phys. Lett. B **362**, 29, 1995

T.S. Biró: *Chaotic Dynamics of SU(2) Gauge Fields in the Presence of Static Charges*, Chaos, Solitons & Fractals, **7**, 645, 1996

T.S. Biró and M.H. Thoma: *Damping Rate and Lyapunov Exponent of Higgs Field at High Temperature*, Phys. Rev. D **54**, 3465, 1996

J.P. Blaizot and E. Iancu: Phys.Rev.Lett **72**, 3317, 1994 and Phys.Lett. B **326**, 138, 1994

C. Greiner and B. Müller: *Classical Fields Near Thermal Equilibrium*, accepted in Phys.Rev.D, hep-th/9605048

U. Heinz, C.R. Hu, S. Leupold, S.G. Matinyan and B. Muller: *Thermalisation and Lyapunov Exponents in the Yang-Mills-Higgs Theory*, Duke-TH-96-129,Phys.Rev.D in press

A.G. Lavkin: *Quark Degree of Freedom of Hadronic Matter and Turbulent Dynamics of Yang-Mills Fields*, Physics of Atomic Nuclei, **59**, 898, 1996

S.G. Matinyan and B. Müller: *Quantum Fluctuations and Dynamical Chaos: an effective potential approach*, Submitted to Found. Phys., hep-th/9610233

Effects of Chaos in Quantum Lattice Systems

F. V. Kusmartsev[1] and K. E. Kürten[2]

[1] L.D. Landau Institute for Theoretical Physics, Moscow,117940, Russia
    Department of Physics, Loughborough University, Loughboro', Leisc., LE11 3TU
[2] Institut für Experimentalphysik, Universität Wien, Boltzmanngasse 5, A-1090 Wien

Abstract. We discuss several quantum lattice systems and show that there may arise chaotic structures in the form of incommensurate or irregular quantum states. As a first example we consider a tight-binding model in which a single electron is strongly coupled with phonons on a one-dimensional (1D) chain of atoms. In the adiabatic approximation the system is described by a discrete nonlinear Schrödinger equation. We have reformulated this equation to the form of a two-dimensional (2D) mapping. By doing this we may investigate a quantum problem in terms usually applied to classical nonlinear dynamic problems. We find three types of solutions: periodic, quasiperiodic and chaotic. The first one are periodic solutions associated with localized deformations of the lattice, like Peierls' charge-density waves or lattice solitons. In the two latter cases the periodicity of solitons breaks down, i.e., the distance between two nearest solitons deviates slightly from the period, that changes randomly, i.e. is unpredictable and therefore chaotic. Thus, we show that the wave function of an electron on a deformable lattice may exhibit incommensurate and irregular structures, analogous to structures arising in classical chaos. We also discuss some other many-body systems including a Kagomé antiferromagnet, where the ground state may have incommensurate structure.

1   Introduction

Classical chaos is a well established phenomenon that we meet everywhere. In brief, the appearance of chaos in dynamical systems means that two slightly different initial conditions correspond to two qualitatively different trajectories in time and/or space such that the long-time behavior of the chaotic system is unpredictable. In quantum mechanics boundary conditions instead of initial conditions play the main role. The question may be put forward if it would be possible in quantum mechanics that a situation analogous to the one in classical chaos arises. To be more precise, is it possible that two slightly different boundary conditions may correspond to two qualitatively different wave functions? Such a dependence on boundary conditions may remind at the classical chaotic dependence on initial conditions.

A possible way to study the quantum analog is to consider a hopping motion of an electron in a solid from one atom to another. Such quantum effects are naturally discrete due to the atomic structure. Thus, one possible system is a single electron motion in a deformable crystal. It is usually described by some kind of tight-binding model. We consider the simplest case of a deformable $1D$ chain of atoms (Callaway 1976), (Ashcroft and Mermin 1979). It is very important to note that in the tight-binding model the motion of the electrons is discrete, since it is assumed that the electron is hopping from one atom to the next. The tight-binding model is still the best approximation for the electron motion in crystals. In very few cases one may introduce a continuum approximation for the tight-binding model in which the motion of the electrons looks similar to that in vacuum. In this case the electron will get an effective mass that is associated with the bandwidth of the original tight-binding model. However, in this case the results may make sense only if the energy values of the electrons are close to the bottom or near the top of the electron band for semiconductors and near the Fermi energy for metals (Ashcroft and Mermin 1979), (Callaway 1976).

Although the Schrödinger equation for the single electron is discrete in the system considered it is still a simple linear equation. However if we take into account the lattice deformations it becomes nontrivial. Since the electron mass is much smaller than the mass of the atoms, we may work in the adiabatic approximation. In this case one may assume that the atoms are moving very slowly and that their kinetic energy can be neglected (Ashcroft and Mermin 1979), (Callaway 1976), (Pekar 1954). Thus the problem is reduced to a stationary one and the Schrödinger equation describes the polaron effect (Pekar 1954), (Rashba 1957), (Holstein 1959) in terms of a cubic nonlinearity. Depending on the type of phonons participating in the effect it has also other names such as "deformon" for acoustic phonons or "condensons" associated with phase transitions. Without addressing a specific case we discuss below a general case and call it a polaron.

2 Classical polarons

The existence of a polaron, i.e. an electron localized (or tightly bound) by the lattice deformation in a crystal was first suggested by Landau in 1933. A first theory of polarons was developed by Pekar (Pekar 1954). Since that time there was much theoretical work on small polarons (Rashba 1957), (Holstein 1959), despite of scarce experimental evidence (Emin 1985).

It has been shown (Rashba 1957), (Rashba 1982), (Kusmartsev and Rashba 1984) that in one-dimensional systems there is a lattice instability, which may give rise to polarons. This is always the case if the density of the electrons is small, while at high densities the polarons may not be created at all, although the lattice instability still exists. For example, for a half-filled band the lattice instability is a Peierls' instability giving rise to charge-density waves and a gap in the spectrum. This phenomenon of the lattice instability occurs both in polar and in nonpolar crystals. We now consider a low-density system where the polarons are only created due to the interaction of the electron with nonpolar phonons, described by the so-called Rashba-Holstein model (Rashba 1957), (Holstein 1959).

The continuum limit of the tight-binding model was first considered by Rashba (Rashba 1957). He studied the equation

$$\frac{\partial^2 \Psi}{\partial x^2} + E\Psi(x) + \frac{D^2}{K}\Psi^3(x) = 0 \qquad (1)$$

and found a single soliton or a self-trapped electron solution in the form $\Psi(x) \sim 1/cosh(ax)$, where $\Psi(x)$ is an envelope of the Bloch wave functions, D and K are the constants of the deformation potential and the elastic modulus, respectively. Notice that Eq. (1) is a special case of a generic equation arising in the study of nonlinear waves

$$\frac{\delta^2 u}{\delta t^2} = \gamma^2 \frac{\delta^2 u}{\delta x^2} + \alpha_1 u(x,t) + \alpha_3 u^3(x,t) \quad . \tag{2}$$

This energy-conserving model shows a variety of spatio-temporal complexity, in particular, *space-time chaos* can be observed (Christov and Nicolis 1996). If only the time or spatial derivative is kept, Eq. (2) describes the motion of the nonforced Duffing oscillator (Duffing 1918) and is equivalent to the continuum limit of our tight-binding model. In contrast to the *forced* Duffing oscillator this model only exhibits regular motion. Note that the Rashba-Holstein model (Rashba 1957), (Holstein 1959) also incorporates *time-dependence* and thus includes both, the atomic and the electron dynamics.

Taking into account the discrete electron motion in solids, the time-independent Schrödinger equation, describing an electron interacting with phonons, is the discrete counterpart of Eq. (1) (Ashcroft and Mermin 1979), (Callaway 1976). It takes the form (Rashba 1957), (Holstein 1959)

$$-\psi_{n+1} - \psi_{n-1} + 2\psi_n + DQ_n\psi_n = E\psi_n \quad , \quad n = -L, ..., -1, 0, 1, .., L. \tag{3}$$

This equation is defined on a finite linear chain (with ψ_{L+1} and $\psi_{-(L+1)}$ set zero) or on a ring consisting of $2L$ atoms. Due to the reflection symmetry of the chain, both for the ring geometry and for the chain, the sites L and $-L$ are equivalent.

Here, ψ_n is the amplitude of the wave function for an electron located at atom n. The deformation potential DQ_n is due to the phonons and the lattice has 2L sites. It has the same form independent of the type of phonons. For acoustic phonons $Q_n = \sum_\alpha (U_n^\alpha - U_{n-1}^\alpha)$, where $\mathbf{U}_n$ is the lattice-deformation vector, while for optical phonons Q_n is the configuration coordinate. The probability to find an electron on atom n is equal to $|\psi_n^2|$. In the adiabatic approximation, where the atomic lattice is only slowly deformed (Pekar 1954), the deformation Q_n is represented by the square of the wave function, $Q_n = -D/K \mid \psi_n \mid^2$, where K is the elastic modulus (Rashba 1957), (Holstein 1959), (Emin 1985), (Rashba 1982), (Kusmartsev and Rashba 1984). The origin of this slow motion is the small ratio of the masses, $m_e/M_{atom} << .001$. The parameters D, K, E are expressed in units of the bandwidth of the electron band t. Inserting $Q_n(\psi_n)$ in Eq.(3) gives the discrete form of the nonlinear Schrödinger equation

$$-\psi_{n+1} - \psi_{n-1} + (2-E)\psi_n - C \mid \psi_n \mid^2 \psi_n = 0 \quad , \quad n = -L, ..., -1, 0, 1, 2, ..., L \tag{4}$$

with $C = D^2/K$. The complete set of solutions of the Schrödinger equation, consisting of $L+1$ independent equations for ψ_n , $n = -L, ..., -1, 0, 1, 2, ..., L$ is obtained by adopting reflection symmetry, $\psi_n = \psi_{-n}$. Thus, the periodic boundary conditions for the ring geometry are automatically satisfied as $\psi_{-L} = \psi_L$. The sites L and $-L$ are identical and this reduces the problem to the analysis of a wave function associated with only $2L$ sites. For simplicity, we discuss only real solutions and therefore we make the substitution $\mid \psi_n^2 \mid = \psi_n^2$ in the Schrödinger equation. In particular, for site $n = 0$ we have

$$-2\psi_1 + (2 - E)\psi_0 - C\psi_0^3 = 0 \quad . \tag{5}$$

In the case of a periodic continuation in space we have to take care of the additional boundary condition for site L,

$$-2\psi_{L-1} + (2 - E)\psi_L - C\psi_L^3 = 0 \quad . \tag{6}$$

For illustration let us first consider the simple case of an electron located on only *two* sites, i.e. $L = 1$. We have two nonlinear equations (5) and (6),

$$2\psi_1 - 2\psi_2 - C\psi_1^3 = E\psi_1 \quad , \tag{7}$$

$$-2\psi_1 + 2\psi_2 - C\psi_2^3 = E\psi_2 \quad , \tag{8}$$

which can be easily solved. Two eigenvalues $E = -C/2$ and $E = 4 - C/2$ correspond to the eigenvectors $(\psi_1, \psi_2) = \pm(1/\sqrt{2}, 1/\sqrt{2})$ and $(\psi_1, \psi_2) = \pm(1/\sqrt{2}, -1/\sqrt{2})$, respectively. Each of these states is two-fold degenerate. Note, that the eigenvectors do not depend on the parameter C, and, moreover the probabilities ψ_n^2 of the electron to be located on each site are equal ($\sim 1/2$). Therefore, the motion of the electron associated with these eigenvalues may be considered periodic with period one, in some arbitrary units. The hopping of the electron from one site to the other one is therefore regular. On the other hand, when the coupling constant C increases there arises the other state having the eigenvalue $E = 2 - C$, which is now four-fold degenerate. The components of the corresponding eigenvectors have the form

$$\psi_1 \to \pm \frac{\sqrt{1+a}}{\sqrt{2}}, \tag{9}$$

$$\psi_2 \to \pm \frac{\sqrt{1-a}}{\sqrt{2}}, \tag{10}$$

with $a = \sqrt{C^2 - 16}/C$. The probabilities of the electron to be located on site 1 and on site 2 are now different. In other words the symmetry between the site 1 and the site 2 is broken and the electron prefers to stay on one site longer than on the other one. Therefore, there is no regular single-period hopping and the total period is two. This localized state corresponds to the creation of a polaron associated with the localization of an electron on one of the two sites. In this toy-model we see the creation of two qualitatively different quantum states. For larger values of L, when the number of lattice sites increases, states associated with many different life-times of the electrons on the individual sites may occur.

As we will see below, a quantum state arises, where the probabilities of the electron to occupy different sites will be very irregular, i.e. chaotic.

In fact, the nonlinear set of equations (4) describing a system consisting of $2L$ sites allows up to 3^{L+1} solutions which - depending on the control parameter E - might be all different. There may exist complex solutions as well which are ignored at this stage. Hence we also have 3^{L+1} possible sets for the wave function ψ_n with $n = 0, \pm 1, ..., \pm L$. Besides the trivial solution (occupation probability zero), one finds homogeneous solutions, where all sites have the same occupation probability,

$$\psi_n^* = \pm\sqrt{-\frac{E}{C}} \quad , \quad n = 0, 1, ..., L \qquad (11)$$

and

$$\psi_n^* = \pm\sqrt{\frac{4-E}{C}} \quad , \quad n = 0, 1, .., L \quad . \qquad (12)$$

One might now be interested in the influence caused by small fluctuations in the initial- or boundary-conditions . For this purpose we first cast the second-order difference equation (4) into the form of a first-order 2D mapping. Introducing the differences $Z_{n+1} = \psi_{n+1} - \psi_n$ we have

$$Z_{n+1} = Z_n - E\psi_n - C\psi_n^3 \quad \text{and} \quad \psi_{n+1} = \psi_n + Z_n - E\psi_n - C\psi_n^3 \quad . \qquad (13)$$

The initial conditions Z_0 and ψ_0 still have to be defined.

For sufficiently large integers n the pair $(\psi^*, Z^*) = (\pm\sqrt{-\frac{E}{C}}, 0)$ can be considered as a "fixed point" of the evolution of Eq. (13) in space. (Solution (12) corresponds to a wave function with space-period *two*.) Linearizing Eq. (13) around the fixed point, Eq. (11), leads to the eigenvalues

$$\lambda_{1/2} = (1 + E) \pm \sqrt{(1+E)^2 - 1}. \qquad (14)$$

Hence, for $E > 0$ the fixed point (ψ^*, Z^*) is unstable ($|\lambda_1| > 1$), while for $-2 < E < 0$ the eigenvalues constitute a complex conjugate pair on the *unit circle*,

$$\lambda_{1/2} = (1 + E) \pm i\sqrt{1 - (1+E)^2}, \qquad (15)$$

independent of the quantity E. (The same analysis with quantitatively the same result can also be performed for the trivial fixed point with occupation probability zero: it is unstable for $E < 0$.) Since the stability analysis only holds for large n we disregard the normalization condition.

Note that the stability behavior does not depend on the parameter C. Since the absolute values of the eigenvalues are unity for all values of $E < 0$ the homogeneous solution is not asymptotically stable. Hence, depending on the choice of the control parameter E *and* the initial condition ψ_0, one might expect either periodic or even aperiodic behavior around the fixed point.

3   Iteration of the map

To explore the nature of possible solutions we first have to find suitable orbits of the Schrödinger equation (4). For small values of L one could try to solve the set of $L+1$ nonlinear equations directly (see section 2). Alternatively, we can follow the lines of methods used in the studies of nonlinear dynamical systems: we can iterate Eq. (4) or what amounts to the same, we iterate the $2D$-map (13).

To start the iteration procedure we prescribe an initial value ψ_0. Due to the symmetry with respect to $n = 0$, quantity ψ_1 has to be taken from Eq. (5). According to Eq. (13) Z_1 then takes the value $\psi_1 - \psi_0$.

The normalization condition

$$\sum_{n=-L}^{L} \psi_n^2 = 1 \qquad (16)$$

can be satisfied by a suitable rescaling method with the aid of the free parameter C. If one rescales the wave function ψ_n by α and the coupling constant C by $1/\alpha^2$ the Schrödinger equation remains invariant. In other words, the value ψ_0 and the value C are related by a similarity transformation: $\psi_n \to \alpha\psi_n$ and $C \to C/\alpha^2$ (Kusmartsev and Rashba 1984). Therefore we use the scaling parameter α to satisfy the normalization condition. The physical value of the electron-phonon coupling is then $c = C \sum_{n=-L}^{L} \psi_n^2$.

The next step employs the following numerical procedure. We first fix the values E and C and iterate the map L times. For negative integers n we adopt reflection symmetry $\psi_n = \psi_{-n}$. In this way we build up the polaron solution with periodic boundary conditions for a chain with $2L$ sites. It may happen that for certain parameter values of E the iteration procedure is divergent. In the case we consider smaller chains. For the original Schrödinger equation this divergence means that at such values of parameters no solution exists.

4   Incommensurate or irregular polarons

There are various possible criteria for the chaos. Numerically the experimental detection of chaos is based on data series in time or space. Phase portraits, or Poincaré maps, where the signal of one variable is plotted against the signal of another variable, indicate chaotic behavior if they are characterized by space-filling points in the two-dimensional plane. For our quantum mechanical problem we take the space trajectory (ψ_n, ψ_{n+1}), $n = 0, 1, ..., L - 1$ that serves as a suitable Poincaré map. Note however that in our problem L is finite and a reliable prediction demands reasonably large data series, i. e. large values of L. Hence we stress again that we intend to detect signatures of "quantum chaos" by means of statistical properties of the discrete nature of the wave function.

We have iterated the map (13) for various values of the energy E. We find that at some fixed initial conditions for the value $E > 0$ there are ordinary solutions for free electrons which are nothing but plane-wave solutions. The polaron non-linear solutions arise for $E < 0$, corresponding to the region of the continuum limit (Rashba 1957), (Rashba 1982). Figs. 1 - 5 provide a selection of

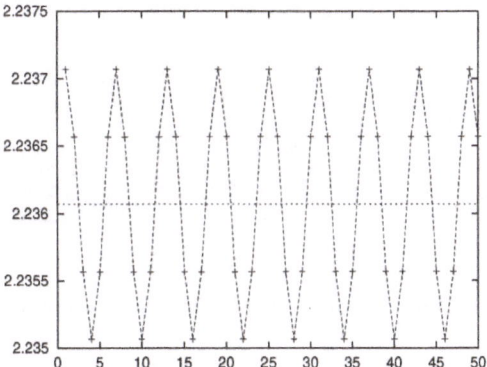

Fig. 1. Dependence of the wave function on the site number $\Delta\psi_0 = 0.001$.

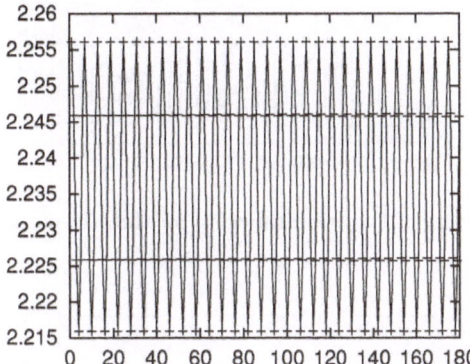

Fig. 2. Dependence of the wave function on the site number $\Delta\psi_0 = 0.02$.

computer experiments, in both ordered and chaotic regions of the wave function for fixed parameters $E = -0.5$ and $C = 0.1$. The corresponding Poincaré maps are presented in Figs. 6 - 7. Note that only the *initial* conditions are subject to variation. They are chosen as

$$\psi_0 = \psi^* + \Delta\psi_0 = \sqrt{5} + \Delta\psi_0.$$

This numerical experiment also gives us information about the stability of the wave function with constant occupation probability.

184

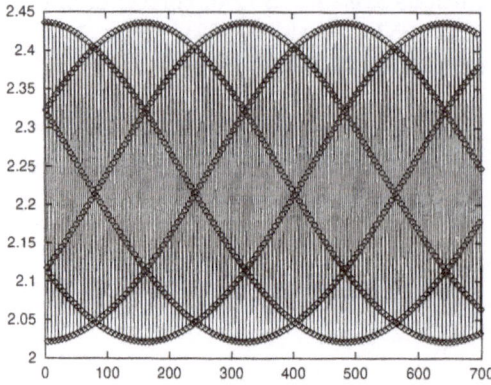

Fig. 3. Dependence of the wave function on the site number $\Delta\psi_0 = 0.2$.

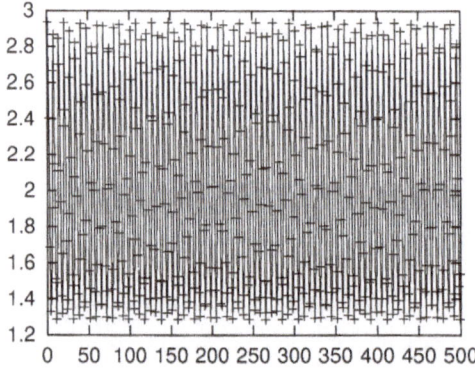

Fig. 4. Dependence of the wave function on the site number $\Delta\psi_0 = 0.7$.

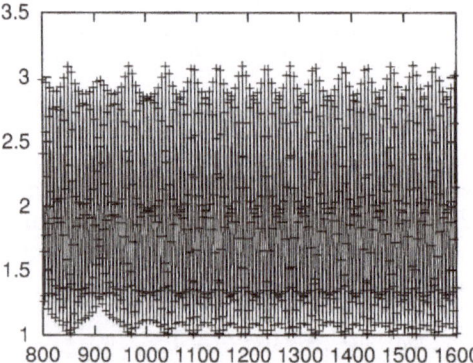

Fig. 5. Dependence of the wave function on the site number $\Delta\psi_0 = 0.743$.

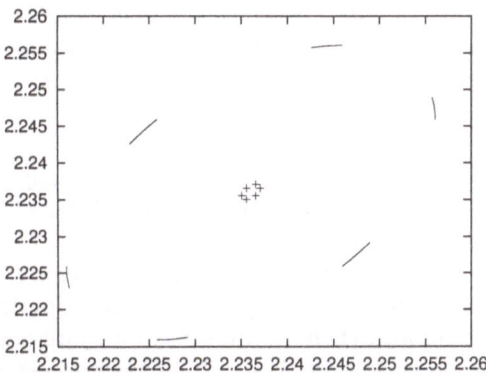

Fig. 6. Poincaré map of the trajectories given in figure 1 and 2 corresponding to figures 3, 4, and 5, respectively (from inside out).

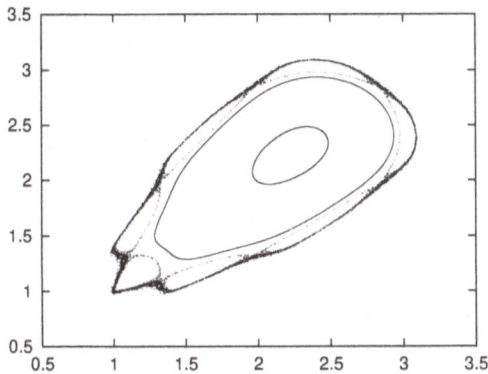

Fig. 7. Poincaré map of the trajectories given in Figs. 3, 4, 5 (from inside out).

As predicted by the linear stability analysis (section 2) the iteration procedure generates oscillating wave functions in space. Fig. 1 depicts a period **6** solution of the wave function for $\Delta\psi_0 = 0.01$. The Poincaré map consists only of **6** points in the two-dimensional plane. This solution represents a generalization of the period two solution for our two-sites model. Here, the electron slightly prefers to sit on each $6 - th$ site. And since the lattice deformation is proportional to ψ_n^2, each 6-th site is more deformed. Such regular deformation gives rise to a superlattice of deformations or a superlattice of solitons. Note that the choice $\Delta\psi_0 = 0$ would yield a constant solution, where all sites of the lattice are homogeneously deformed. Increasing $\Delta\psi_0$, the deviation from $\psi*$, the period **6** solution looses its stability. Then the superlattice breaks down, although this break-down is still very regular. This means that the deviation from the superlattice is monotonically increased with the number of the site n. Thus, the periodicity of the wave function breaks down, while the lattice deformations still approximately preserve some quasiperiodicity. For a finite-size system a small deviation from the period cannot be detected and we still observe a lattice of solitons.

Further inspection of Fig. 2 and especially the Poincaré map Fig. 6 shows clearly that the solution is not periodic on the lattice any more. With increasing L the points visited in phase space become more and more dense although no point is visited twice.

One may analyze the quasiperiodicity of the envelope. Since the deviation is almost regular this envelope creates periodic or quasiperiodic structure as well. If our system were infinite then we may have a system with two periods, the ratio of which may be very large. For a finite system, especially of small size, it may be considered as belonging to the previous class.

This quasiperiodic (Kürten and Nicolis 1997) solution may be of two types as well. It depends on the ratio of these two periods. If it is rational then there are two commensurate superlattices. On the other hand, if the ratio of these two periods is irrational one may have an incommensurate structure. However, the incommensurability is still very regular. To summarize: for small $\Delta\psi_0$ the phase portraits are described by a closed elliptical invariant curve. With increasing $\Delta\psi_0$ the radii increase until at $\Delta\psi_0 = 0.7$ the elliptical shapes undergo deformations. Eventually, we find a smooth transition to weak chaotic behavior at $\Delta\psi_0 = 0.743$.

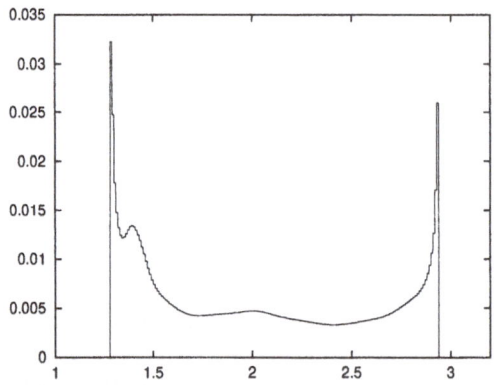

Fig. 8. Probability-density for $\Delta\psi_0 = 0.7$

In marked contrast to the sharply peaked probability-density of a *periodic* wave function Figs. 8 and 9 shows the probability density in the nonperiodic regimes. In the quasiperiodic regime the density is continuous (Fig. 8), whereas in the chaotic regime the piecewise discontinuity might indicate fractal structures of the wave function (fig. 9). Fractal structures are indeed intimately related to chaos and might be a manifestation of the appearance of chaos in quantum mechanics. However, the question of fractality at this stage of our investigations

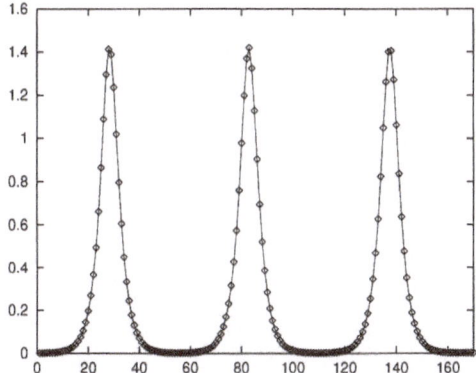

Fig. 9. Probability-density for $\Delta\psi_0 = 0.743$

has still a suggestive character and has to be carefully examined. Note that the quasiperiodic orbits are dense in the plane. That is, changing slightly the initial condition, also the Poincaré map will be slightly different. Since slight variations in the initial conditions will always lead to slightly different orbits, they are not attractors.

For physical illustrations it is more appropriate to work with the control parameter E instead of the initial value ψ_0, since the electron energy E is experimentally accessible. For that reason we perform a second series of computer experiments, fix the initial condition $\psi_0 = 0.001$, and vary the physical parameter E. The associated space evolution of the wave function is presented in Figs. 10 - 12, whereas illustrative Poincaré maps are juxtaposed in Fig. 13.

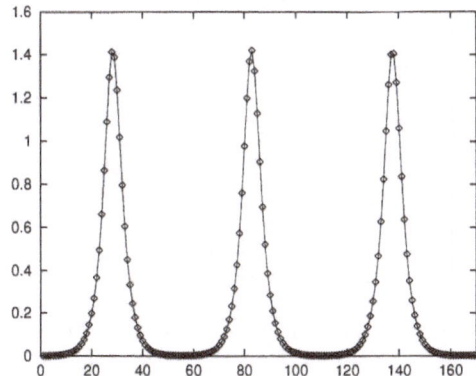

Fig. 10. The dependence of the wave function on the site number, $E = -0.1$.

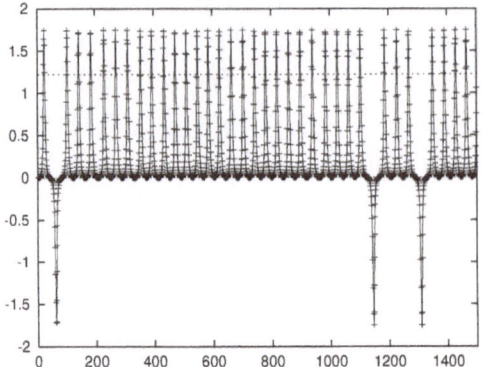

Fig. 11. The dependence of the wave function on the site number, $E = -0.15$.

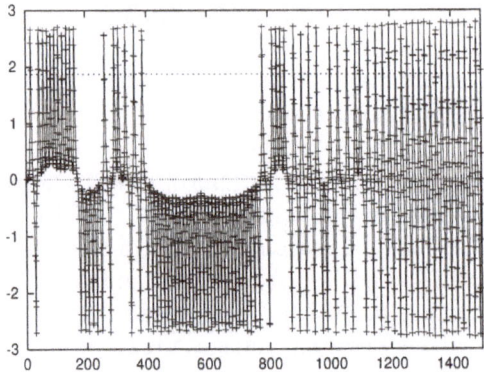

Fig. 12. The dependence of the wave function on the site number, $E = -0.35$.

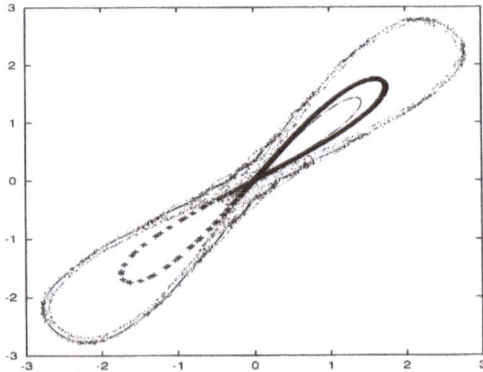

Fig. 13. Poincaré maps of the trajectories given in figure 10 - 12 (from inside out).

For the values $E = -0.1$ and $E = -0.15$ we find quasiperiodic behavior, where the phase trajectories wander between two manifolds close to the fixed points $\psi^* = \pm\sqrt{-\frac{E}{C}}$. For increasing absolute values of E we eventually observe space filling points indicating transition to chaos.

5 Chaos in many-body systems or creation of incommensurate Wigner crystals

The approach discussed above and the way to find regular and irregular structures may be suitably generalized for an analogous study of many-body systems. In the former example the state vector may be represented by

$$|1\rangle = \sum_n \psi_n a_n^\dagger |0\rangle , \qquad (17)$$

where $a_n^\dagger$ is a single-particle creation operator on site n. The many-body generalization of the state vector is

$$|1\rangle = \sum_{n_1,\ldots,n_k} A_{n_1,\ldots,n_k} a_{n_1}^\dagger \ldots a_{n_k}^\dagger |0\rangle . \qquad (18)$$

Here, the role of the amplitude of the single-particle wave function ψ_n is played by the configuration amplitude $A_{n_1,\ldots,n_k}$. The conjugate variable Z_n of the single-particle problem acting on the state vector $|0\rangle$ is now replaced by an operator as

$$P|1\rangle = \sum_n (\psi_{n+1} - \psi_n) a_n^\dagger |0\rangle = \sum_n Z_n a_n^\dagger |0\rangle, \qquad (19)$$

where the operator P has the form

$$\sum_n (a_n^\dagger a_{n+1} - a_n^\dagger a_n). \qquad (20)$$

In doing this we project the Hilbert" phase space" onto a plane generating the Poincaré map for the quantum state $\{\psi_n, Z_n\}$. In the many-body problem we may adopt an analogous procedure, employing a current operator I instead of the operator P. More specifically, we apply the operator I on the state vector

$$|1\rangle = \sum_{n_1,\ldots,n_k} A_{n_1,\ldots,n_k} a_{n_1}^\dagger \ldots a_{n_k}^\dagger |0\rangle \qquad (21)$$

and get

$$I \sum_{n_1,\ldots,n_k} A_{n_1,\ldots,n_k} a_{n_1}^\dagger \ldots a_{n_k}^\dagger |0\rangle = \sum_{n_1,\ldots,n_k} B_{n_1,\ldots,n_k} a_{n_1}^\dagger \ldots a_{n_k}^\dagger |0\rangle \quad . \qquad (22)$$

Therewith, our many-body problem is now represented by the set $\{A_{n_1,\ldots,n_k}, B_{n_1,\ldots,n_k}\}$. This is an appropriate $N-$dimensional generalization of the Poincaré map for the quantum state discussed and in fact a projection of the multiparticle Hilbert-space onto the plane. If this set is nontrivial it may indicate the appearance of irregular effects in the many-body system.

To illustrate these ideas one may study a simple example of electrons located on a square $L \times L$ lattice. The single-particle wave function is a plane wave,

$$\psi_{n,m} = \exp(i\mathbf{k} \cdot \mathbf{r})/L \qquad (23)$$

where the vectors $\mathbf{k} = (2\pi l_1/L, 2\pi l_2/L)$ and $\mathbf{r} = (na, ma)$ with l_1 and l_2 are some integers. Du to the choice of vector $\mathbf{k}$ it is commensurate with the lattice and irregular structures cannot arise. The same may be true for the many-body system, where the many-body wave function $A_{n1,...,nk}$ is a Slater-determinant consisting of plane waves. Since each plane wave is commensurate with the lattice, the many-body wave function $A_{n1,...,nk}$ built out of plane waves is also commensurate with the lattice. Any vector $B_{n1,...,nk}$ generated from that one will be commensurate with the lattice as well. Again, chaotic structures do not arise.

The situation is different if we take into account Coulomb interactions between the electrons. At low densities the electrons may be localized on the lattice sites forming a Wigner crystal. The size of the unit-cell and the lattice constant of the Wigner crystal depend on the electron density. The Wigner lattice and the original lattice where the electrons are located may not be commensurate, i.e. the ratio of their periods cannot be represented as a rational number. One may expect the appearance of incommensurate or irregular solitons (Luther, Timonen, and Pokrovsky 1980). Consequently, these solitons (Luther, Timonen, and Pokrovsky 1980) give rise to a chaotic type of the Poincaré map of the ground state.

The present discussion is limited to systems with a finite number of particles. In the thermodynamic limit one can for example adopt correlated-basis-function theory (Gernoth, et. al. 1994), (Ristig and Kim 1996), where the pair distribution function $g(r)$ may reflect the presence of the incommensurate state. The type of these correlation functions should depend on the nature of the possible chaotic structure. For example, for an irregular Wigner crystal, we expect that the many-body wave function is fractal.

6   Chaos on a Kagomé lattice

We consider a spin-lattice system, say an anti-ferromagnetic phase, on a Kagomé lattice. As usual, the system is described by the Heisenberg Hamiltonian

$$H_{KA} = J \sum_{<i,j>} S_i S_j \quad , \qquad (24)$$

with positive exchange-constant $J > 0$ and the spin operators S_i located on the Kagomé lattice sites. The nature of the ground state of this system is still a subject of debate. The classical version of the model allows numerous continuous symmetries. This provides a connection between a Kagomé lattice and an XY model, that has $O(2)$ symmetry. On the other hand, spin configurations on a single triangle of the Kagomé lattice can induce frustration. The frustration results in the appearance of vortices or other non-linear effects like solitons. This, in turn may create a commensurate or an incommensurate lattice, even in the ground state. To give arguments in support of this expectation let us proceed as follows.

In the previous single-particle example the configuration space consisted of the single-particle basis

$$|n\rangle = a_n^\dagger |0\rangle \quad . \tag{25}$$

The amplitude of each configuration is defined by the single-atom wave function ψ_n. In analogy to this basis $|n\rangle$ we introduce the configuration space of lattice spins by

$$|Q_n\rangle = S_1^z S_2^z S_3^z S_4^z \cdots S_n^z |0\rangle \quad . \tag{26}$$

The model may be diagonalized, for example, by employing the Lancsos' algorithm. The result may be represented by a set of eigenvalues $\{E\}$. For each of these eigenvalues we have a set of amplitudes $\{\Psi_n\}$. Each of them is associated with a spin configuration Q_n. In a next step we must find a quantity on the Kagomé lattice that is equivalent to the quantity Z_n in the single-particle problem.

In analogy to the operator P that generates the amplitude Z_n in the single-particle basis we may introduce the density of a chiral current, that is associated with the permutations of three spins (Bernu, Lhuillier, Lecheminant, and Zindzingre 1997), $C_n = P_{n,n-1}P_{n+1,n} - P_{n,n+1}P_{n-1,n}$, where $P_{n,m}$ is a permutation of spins located on site n and m. Acting wth the operator P on the state vector with amplitudes ψ_n we generate the set of amplitudes C_n associated with the original configuration Q_n. By this construction we get the Poincaré map for small Kagomé clusters ($N = 21$ sites) and the lowest eigenvalue indicates the existence of chaotic structures. To present some results it is more convenient to separate the amplitude and the phase of the wave function. This yields four different sets

$$\{vm = \text{Ampl}(\Psi_n), vtm = \text{Ampl}(C_n)\} \, , \tag{27}$$

$$\{vm = \text{Ampl}(\Psi_n), vta = \text{Phase}(C_n)\} \, , \tag{28}$$

$$\{va = \text{Phase}(\Psi_n), vtm = \text{Ampl}(C_n)\} \, , \tag{29}$$

$$\{va = \text{Phase}(\Psi_n), vta = \text{Phase}(C_n)\} \, . \tag{30}$$

Numerical results on these dependencies are presented in Fig. 14, respectively. (taken from calculations by B. Bernu, et al.) These phase portraits are an indication for the existence of chaotic structures in the ground state of the Kagomé Antiferromagnet. Correspondingly, the order parameter is also irregular.

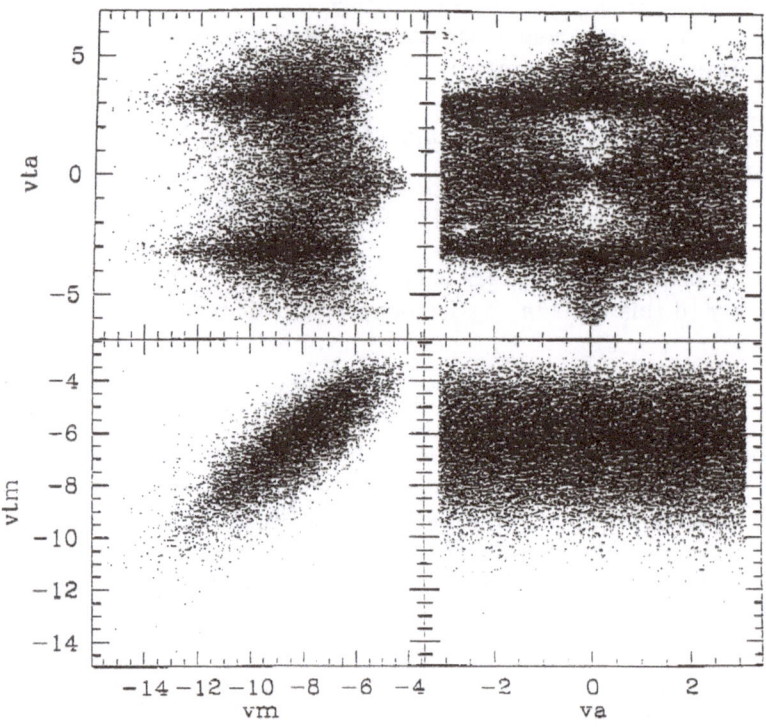

Fig. 14. Poincaré map $\{Z_n, \psi_n\}$ for the ground state of the Heisenberg antiferromagnet on a Kagomé lattice. These maps indicate the incommensurate structures of the ground state .

7 Irregular and incommensurate ground states

In general, lattice symmetry is very important for describing the structure of the ground state. A typical example is the drastic difference between the behavior of Heisenberg spins on square- and triangular-lattices. The difference is caused by the frustration induced on the triangular plaquettes. This frustration may be effectively described by introducing an appropriate gauge field. An interesting example where frustration may be described by a gauge field is the Kagomé antiferromagnet. The classical spins on a Kagomé lattice have two different configurations allowed by a continuous symmetry, such as $U(1)$ multiplied by L or L^2 in the first and the second case, respectively. The first configuration may be called a "$q = 0$ state", the second configuration a "$\sqrt{3} \times \sqrt{3}$ state". The appearance of these symmetries means that the classical state is strongly degenerate. This degeneracy corresponds to the possibility of free spin rotations without a change of energy. In the configuration $\sqrt{3} \times \sqrt{3}$ one may rotate 6 spins located around a hexagon. In the configuration $q = 0$ one may rotate the spins located around the straight lines along the main directions of the Kagomé lattice without any restrictions. By the slow (adiabatic) rotation the total change of the gauge phase is equal to $+2\pi$, i.e., $\sum \phi_{i,j} = +2\pi$, where the phase $\phi_{i,j}$ is the phase change on the nearest sites. We may say that a vortex is created. The local $U(1)$ symmetry is therefore responsible for the creation of vortices and anti-vortices, or the $U(1)$ symmetry is giving rise to chiral degrees of freedom. These vortices are strongly (logarithmically) interacting with each other. Due to this attraction the vortices and the anti-vortices are paired or annihilated in pairs. The pairing generates a phase coherence between all hexagons, similar to the coherence effect that occurs in a superconducting or in a superfluid state. The energy gap or the superfluid density and the phase may serve as an order parameter characterizing the phase.

For a quantum Kagomé antiferromagnet the density and the phase are conjugate variables. Thus, if there is phase coherence the density is fluctuating and vice versa. The superfluid density is the density of the "hard core" bosons, i.e. $n_i \sim S_i^z$ (the z-axis projection of the spin orientation). The fluctuating density implies that the spin-degrees of freedom are in a disordered state or in an irregular state, consequently, the correlation function $< S^z(1)S^z(2) >$ is short ranged. We conclude that in the ground state the chiral degrees of freedom are ordered, while the spin-degrees of freedom are disordered. This conclusion is illustrated in Fig. 14. We see that in the ground state all spin configurations are not commensurate with the lattice and moreover create the fractal behavior. Possibly, this irregular incommensurability creates the disordered (spin liquid) ground state (Bernu, Lhuillier, Lecheminant, and Zindzingre 1997).

8 Conclusions

We have shown the formation of chaotic structures in the polaron wave function on a finite-size system. Due to the lattice deformations the electron is localized on different sites with different probability. There are also regular periodic or quasi-periodic solutions where the wave function amplitudes replicate with some period equal to some integer number of lattice constants or create a regular incommensurate structure.. Thus we find chaotic or irregular structures in the lattice wave-function. From a quasiclassical point of view the electron moving in one direction exhibits very irregular hopping from one site to the next. The chaotic behavior in *space* obtained is certainly due to the atomic lattice discretization associated with the atomic structure of solids. The predicted space chaos is robust and should be experimentally observed. To characterize irregular incommensurate quantum ground-states one may use all methods applied in studies of classical chaos. Employing these methods we find two interesting types of ground states. The first one is that the lattice polarons may have a state or create a soliton lattice having a period that is incommensurate with the period of the atomic lattice or may have no fixed period at all. The second one is that the ground state on the Kagomé antiferromagnet is irregular incommensurate. Its chiral degrees of freedom create a superlattice with a period irregular or incommensurate with the spin lattice. The chiral degrees of freedom are associated with the rotations of a group of spins around hexagons or along straight lines of the main directions. Ordered chiral degrees of freedom may correspond to a disordered state of spins. Further investigations are needed to shed light on these very interesting and complex issues.

Acknowledgments

The work was supported in part by the Academy of Science of Finland and by Nordita. KEK acknowledges partial support by the Training and Mobility of Researchers program of the European Commission and by the Poles d'Attraction Interuniversity program of the Belgian federal office of scientific, technical and cultural affairs. We acknowledge fruitful discussions with K. Alekseev, B. Bernu, T. S. Biro, K. Gernoth, J. Hofbauer, C. Lhuillier, P. Lecheminant, E. Manousakis, G. Nicolis, M. L. Ristig, P. Zindzingre, and H.S.Dhillon.

References

Ashcroft N.W. and Mermin N.D., Solid State Physics, (Cornell University, 1976).

Bernu B., Lhuillier C., Lecheminant P., and Zindzingre P., Phys. Rev. Lett. 1997, submitted.

Callaway J., Quantum Theory of the Solid State, (Academic Press, NY, 1976).

Christov C.I. and Nicolis G., Physica A **228**,326 (1996).

Duffing G., Erzwungene Schwingungen bei veränderlicher Eigenfrequenz und ihre technische Bedeutung (Vieweg & Sohn, Braunschweig, 1918).

Emin D, Phys. Today,**35**, 111 (1985); Phys. Rev. Lett. **62**, 1544 (1989).

Gernoth, K.A., Clark, J.W., Senger, G., and Ristig, M.L.: Phys. Rev. B **49**, 15836 (1994).

Holstein H., Ann. Phys. (N. Y.) **8**, 343 (1959).

Kürten K.E. and Nicolis G., Physica A (1997).

Kusmartsev F.V. and Rashba E.I., Sov. Phys.- JETP **59**, 668 (1984).

Luther A., Timonen J. and Pokrovsky V., in *Phase Transitions in Surface Films*, eds. J.G.Dash and J. Ruvalds, (Plenum, New York, 1980).

Pekar S.I., Untersuchungen über die Elektronentheorie der Kristalle, (Akademie Verlag, Berlin, 1954).

Rashba E.I., Opt. Spectr. **2**, 88 (1957).

Rashba E.I., in *Excitons*, ed. by E.I.Rashba and M.D. Sturge , (North-Holland, Amsterdam, 1982) p.543.

Ristig, M.L. and Kim, J.W.: Phys. Rev. B **40**, 6665 (1996).

New Series m: Monographs

Lecture Notes in Physics

For information about Vols. 1–461
please contact your bookseller or Springer-Verlag